Business Ventures in the Former Soviet Union

Negotiation and Protocol
Dos* and *Don'ts

Business Ventures in the Former Soviet Union

Negotiation and Protocol *Dos* and *Don'ts*

Joseph A. Kliger

PennWell Books

PennWell Publishing Company
1421 South Sheridan/P.O. Box 1260
Tulsa, OK 74101

Library of Congress Cataloging-in- Publication Data
Kliger, Joseph A.
Business ventures in the Former Soviet Union : negotiation and protocol, dos and don'ts / Joseph A. Kliger.
p. cm.
Includes bibliographical references and index.
ISBN 0-87814-428-5
1. Negotiation in business--Former Soviet republics. 2. Business etiquette--Former Soviet republics. 3. Investments, American--Former Soviet republics. 4. Former Soviet republics--Commerce--United States. 5. United States--Commerce--Former Soviet republics. 6. Petroleum industry and trade--Former Soviet republics. I. Title.
HD58.6.K59 1994
658.4--dc20 94-40713
CIP

Printed in the United States of America

1 2 3 4 5 98 97 96 95 94

Contents

Foreword vii

Acknowledgements xi

PART 1 Moslem Belt of the Former Soviet Union 1

Introduction 3

1. Kirgizstan: Gold Rush 9
2. Uzbekistan: Wages Talk 47
3. Chechen: Willful Men 71
4. Tatarstan: Old Energy 93
5. Moslem Belt of the Former Soviet Union: Rules, Comments, Advice 105

PART 2 This Is Russia 117

Introduction 119

6. Moscow's Volcanoes 123
7. Saratov: The Province 147
8. Calculations on Western Siberia 157
9. Black Sea: Onshore and Offshore 177
10. This Is Russia: Rules, Comments, Advice 189

Index 195

Foreword

The best way to avoid trouble in Russia is to stay away from this country. However, since you have this book in your hands, you haven't made this decision yet. It could be possible that after reading this book you and your money will never appear in areas of the former Soviet Union (FSU), or you may develop more doubts if you are already involved. I'd suggest that after reading this book you'll become more qualified to make informed decisions.

This book's *propaganda* is rather simple:

- If you wish to engage in business within the FSU, you have a chance of making a profit.
- If you are already involved, don't withdraw. You still have a chance.
- If you had and have doubts, don't worry about missed opportunities. Your doubts were probably justified.

Russia and the republics of the FSU aren't machines for making money. They are countries choking with their own problems, and you are doomed to share some of these problems if you operate there. They are extremely rich countries, and they are doomed to share their riches with you if you share their problems with them. By solving your own problems in the FSU, you'd solve some of their problems too. It might look attractive, but you must remember that these problems aren't easily solved.

In the described regions, you will face problems from the very beginning of your activity, and your American mind will prompt you to call in a number of professionals who could help: lawyers, geologists, economists, and possibly an astrologist. Your American mind will happen to be wrong very often since it is now required to deal with a very different culture. (It is better to say: with a few very different cultures.)

This book—by framing the problems within their unique cultural contexts—will enable you to devise appropriate and successful strategies. Following this method, it is much easier to recognize which features *are* and *are not* important for your business in the FSU.

For example, it will become clear to you why the political instability and those expected and unexpected changes in the future *are not* too important since each subsequent ruler of a certain corner of the FSU will need a foreign presence in his country. However, a seemingly insignificant thing such as saying "good-by" in a proper manner *is* important since you want him to choose *your* presence.

In this book you'll find some more examples as to how to help these key figures in making their choice.

In the summer of 1992 at Lake Issyk Kul in Kirgizstan, I heard Mr. A. Kozyrev, Russian foreign affairs minister, quoting his famous ancestor Mr. A. Gromyko as having said: "Sovietologist? I have them for free. I don't care about their theories, but I would pay them for each advice that they give to their State Department."

These words were recalled to my mind constantly while writing this book.

Sovietologists' theories and advice belong to the times when the FSU needed nothing but enemies. Totalitarianism justifies itself by creating enemies. Of course, it was a difficult issue for an American mind to understand. Thus, sovietologists had to invent some understandable theories for themselves, but those formed an evidently wrong base for issuing qualified advice. It was really difficult to avoid mistakes since the kernel of the conflicts between the United States of America and the FSU was artificial and located in the area of ideology where most Americans didn't have and didn't wish to have any experience.

The FSU and its ideology don't exist anymore.

Modern Russia and the other republics have enemies but don't seek or create them. This is understandable. Now, the conflicts between Russia and the West arise from economic matters and become real conflicts of real interests, which is also understandable.

Americans and Europeans have much experience in solving these kinds of problems among themselves and around the world (Japan, Middle East). But Russians and Russia, Kirgiz and Kirgizstan, and people in the other republics don't have any of this experience, and therefore don't know how to defend their own interests and how to reach, enrich, and share the mutual profit. This is why their reaction to various proposals often has to be overcautious.

The visitors from the West don't understand this, aren't patient enough, and don't have any respect for what is assumed as the historically stipulated ignorance of the local people.

The local people are seeking personal interests in each deal, don't accept fast decisions, are very suspicious on the issue of colonialism, and so on.

But time passes. Both sides have made enough mistakes to learn some of their lessons.

Now, the moment comes to put an end to this competition in incompetence.

Acknowledgements

All the data and information for this book were obtained from sources in the former Soviet Union.

I owe a considerable debt of gratitude to these persons, especially:

Byelorus:

V. N. Kuznetsov, First Deputy Speaker of the Parliament.

R. M. Nourgaleev, General Director of the local geophysical company *Zapneftegeofizika.*

Chechen:

R. Goulburaev, Minister of Oil and Chemical Industry.

Crimea:

V. N. Naidenov, Chief Geologist of the local oil company *Chernomorneftegas.*

Dagestan:

M. Djabraielov, Chief Geologist of the local oil company *Dagneft.*

Kirgizstan:

F. Kulov, Vice-President.

N. Murtazaliev, Chief Geologist of *Kirgizzoloto* (Kirgiz Gold).

B. Shamshiev, famous movie director.

A. Zjumagulov, Prime Minister.

Moscow:

A. F. Ageeva, authors' secretary.

I. S. Alekseev, former Vice-President of *Almazzoloto* (Diamonds and Gold) Corporation.

A. B. Calenkin, former Deputy Minister of Finance of Russian Federation.

P. A. Gerish, Chief of Department of *Gasprom* (Gas Industry).

V. M. Nikitin, Director of VNIPINeft Institute.

S. E. Taslitsky, Deputy Minister of Energy and Fuel of Russian Federation.

N. Y. Uspenskaya, former economic advisor to the Prime Minister of the former Soviet Union.

Saratov:

A. V. Michurin, President of the local geophysical company *Saratovneftegeofizika.*

P. A. Zakharov, Vice-President of *Saratovneftegeofizika.*

Stavropol:

I. I. Kouchougoura, Chief Engineer of the local geophysical company *Stavropolneftegeofizika.*

Tatarstan:

A. Rijanov, Chief Gas and Oil expert of the Cabinet of Ministers of Tatarstan.

Ukraine:

V. I. Oleksuk, Chief Geologist of *Ukrgasprom* (Ukrainian Gas Industry).

Uzbekistan:

S. Najimov, Chairman of the State Committee on Precious Metals.

A. Rakhimov, First Deputy Chairman of the State Concern *Uzbekneftegaz.*

B. Tursunov, Chancellor of the Independent University of Business and Diplomacy.

West Siberia:

V. N. Chepursky, Chief Engineer of *Sibnefteprovod* (Siberian oil pipelines).

G. A. Scherbakov, plenipotentiary of the President of Russia in Tyumen Region.

USA:

Mr. A. Petzet, Editor of the *Oil and Gas Journal* published two of my articles in this well-known magazine. He had never met me, but when I explained the idea for this book by phone, he connected me with Ms. S. R. Sesso the next day. Ms. S. R. Sesso, Acquisitions Manager for PennWell Books, was very enthusiastic about this book after reading a few pages—even though she knew that English wasn't my native language. This current publishing experience has become a pattern for me as to what can be done if you want things to be done. I cannot avoid comparisons: I have published a few books in Moscow and it took three to four years for manuscripts to be approved and published.

My American friends who know Russia and the Russian language—M. Flaxman (President of Grynberg Resources), P. Buckley (Vice-President of Petron), A. Guff (Vice-President of Paine Webber)—read the manuscript or its parts and asked questions. The answers I knew are in the book.

And my wife, Vicky, has corrected almost every sentence in this book. She also had several ideas about what matters would interest the female readers. After discussing her ideas, I think I understood why businesswomen are proportionally more successful than men in the former Soviet Union.

And my nephew, Elly Kliger, a student of philosophy at Cornell University, who tried to persuade me that Russia itself is a philosophy.

Part 1

Moslem Belt of the Former Soviet Union

Introduction

The land known as the Moslem's Belt of the former Soviet Union (FSU) extends from the Altay Mountains across the great deserts and steppes of central Asia westward through the Caspian Sea and along the Caucasus (Fig. 1–1). The land mass is more than 5,000 km long by 2,000 km wide. The Moslem's Belt is populated by the mixed descendants of different ancient tribes of nomads: Scythians, Huns, Khazars, Mongols, Tatars, Turki, etc. Even though their languages, cultures, and habits are different, all became Moslems in the twelfth century. The powerful impact of this great religion changed their civilization but couldn't change their geographical location and, therefore, their historical fortune. They are located next to Russia.

After the disintegration of the FSU in 1991, Kazakhstan, Uzbekistan, Kirgizstan, Azerbaizian, Turkmenistan, and Tajikistan (Fig. 1-1) became independent states; 8 other Moslem nations became republics in Russian Federations. (The Chechen Republic and Tatarstan declared their independence, but neither the FSU nor the UN recognized them.)

Those Russians who claim to be Europeans call this geographical belt fetters of their historical fortune; those Russians who insist that their country is The Continent (and Europe only a peninsula to it) wouldn't call this zone *a belt* either since they feel it is part of Russia's body (not a dress). The brusqueness of the following statement typifies the feelings of those Russians living in the frontier.

> "I'd like to have only one thing in common with them—a tall, tall, tall fence."
>
> (Couchougoura, Chief Engineer, *Stavropolneftegeofizika* [Stavropol Geophysical Survey]; conversation with author, Stavropol, May 1993)

The Moslems would answer that they live in their land, in their own houses, and on their own. This is true. But it is also true that

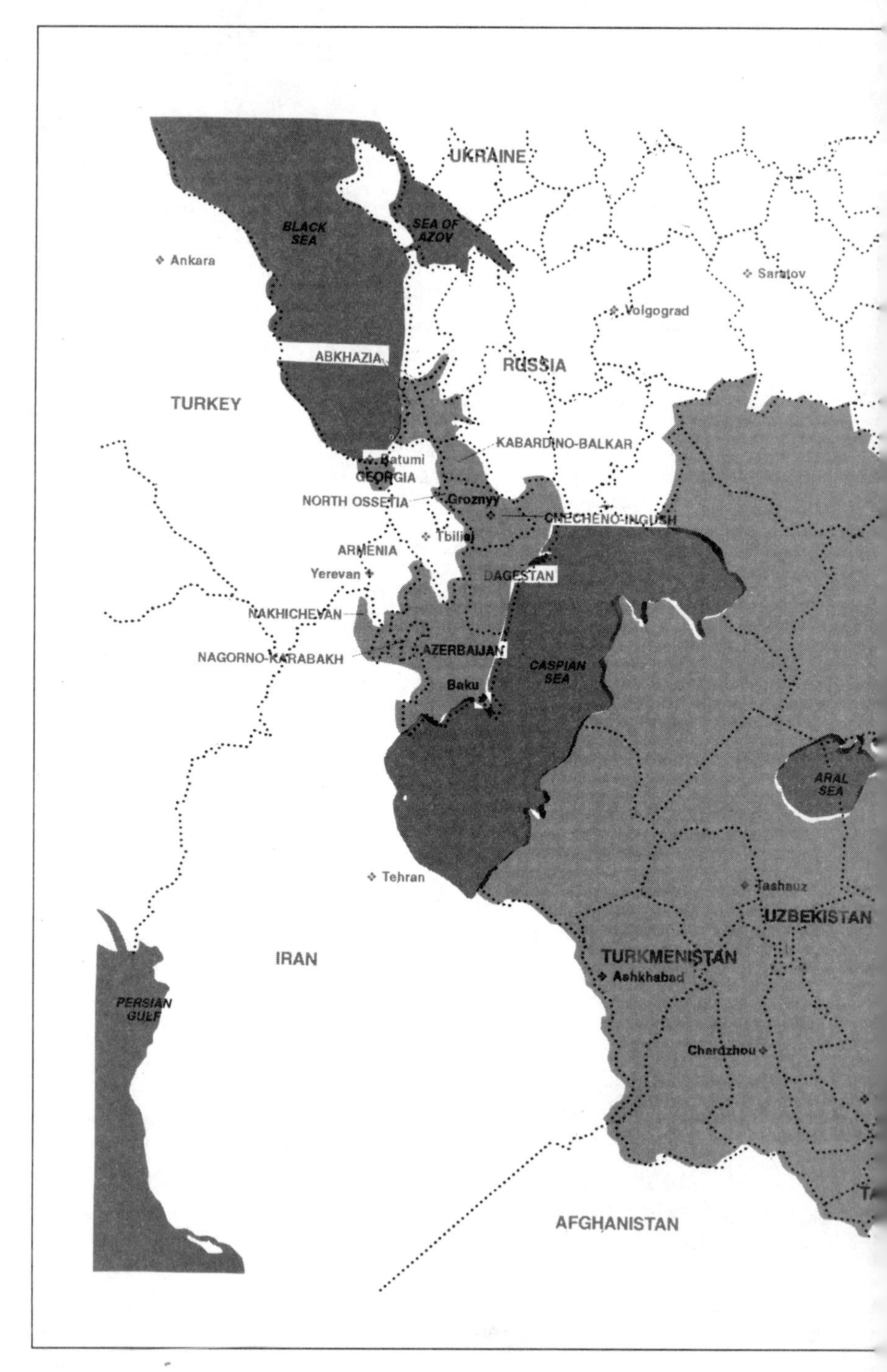

Figure 1–1 Moslem Republics of the Former Soviet Union

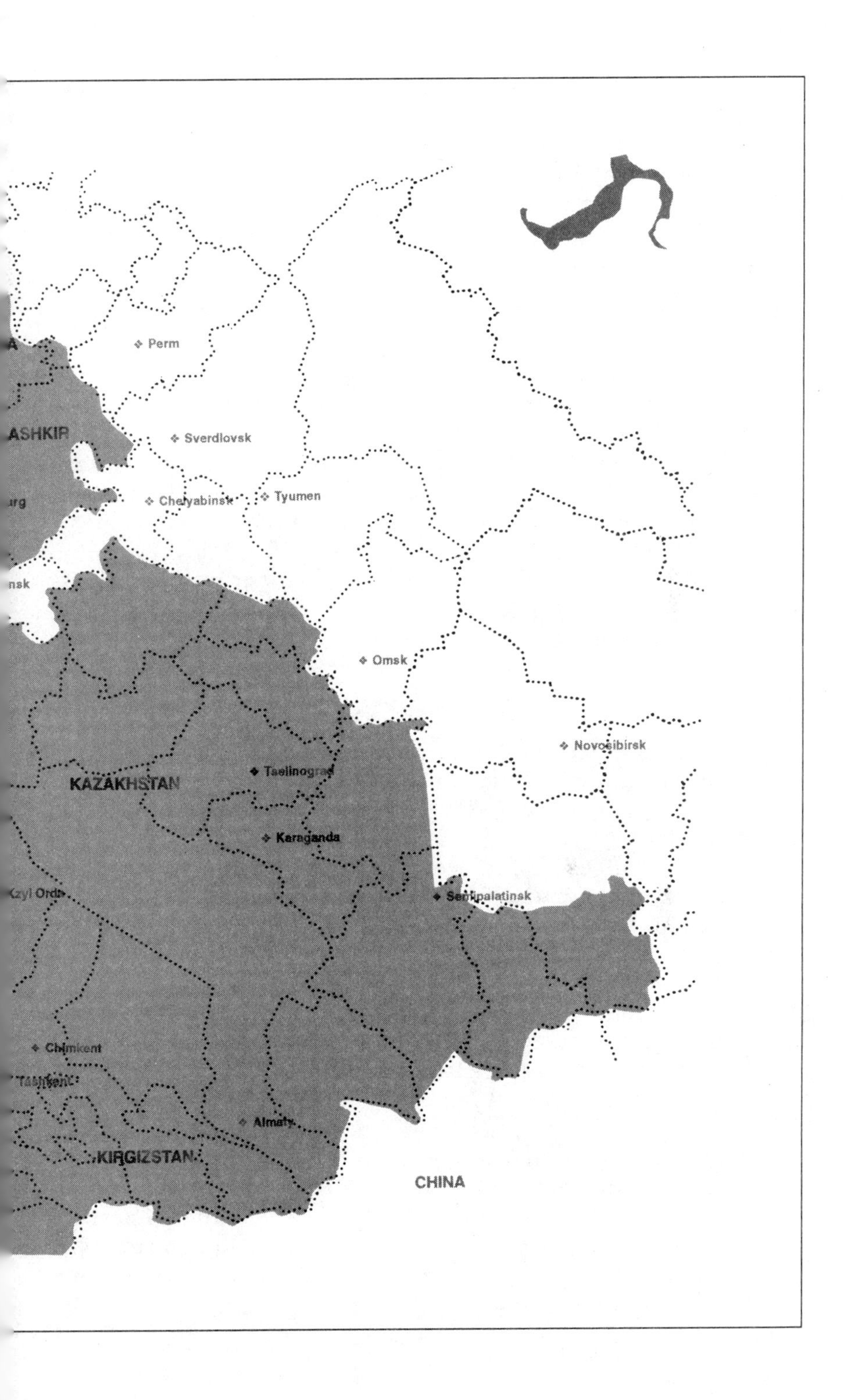
Perm
Sverdlovsk
Chelyabinsk
Tyumen
Omsk
Novosibirsk
KAZAKHSTAN
Tselinograd
Karaganda
Semipalatinsk
Chimkent
Almaty
KIRGIZSTAN
CHINA

they have to check whether or not they can afford a tall fence along their northern frontier.

The Koran says that both are punished already: the one who conquers — for his ill will; the one who is subjugated — for he indulges the evil outbursts.

I don't dare comment on the Koran; I am, however, curious as to whether or not this punishment ends when the relationship changes. Britain was lucky to have its colonies at a distance of thousands of miles and across oceans; Russia does not have this luxury.

Back to the steppes.

The indigenous people are traditionally hospitable but very xenophobic; they are warm friends but ice-cold enemies; they are extremely ambitious, but they are ready to sacrifice any ambitions in the name of the hierarchy (they are exquisitely sensitive to their own position therein); they are complacent and obliging to those in power, but they become willful and even cruel when they are in power (and they understand which role to perform in which circumstance).

You face trouble communicating with these people, considering that they never had direct contact with the West, but only through Russia.

> "My initial feeling during our first meetings was a kind — how to describe this? — a kind of emptiness. . . . Yes, that was it. I felt it and I think they did also."
>
> (T. R. Hutson, First Secretary, U.S. Embassy in Kirgizstan; conversation with author, Bishkek, November 1992)

Here is a classic example of how this lack of experience works in the area of politics. (In the following chapters you'll find many examples related to business.) In October, 1993, U.S. Secretary of State Warren Christopher visited Kazakhstan. "No exchange of gifts," was an American requirement. American diplomacy showed its best intention by implying that the relationship between the two countries is far more important than that between the personalities. But didn't Secretary Christopher's advisers know that the tradition of giving gifts in Oriental diplomacy is older than America, England, and even the Roman Empire? Didn't they know that a single cowboy hat handed to Mr. Nazarbaev[1] was worth more than all of the humanitarian aid of $15 million to the people of Almaty?[2]

1. The President of Kazakhstan.
2. The capital of Kazakhstan.

Each Kazakh could explain this custom. What is the greatest insult to an Oriental host? It is when a servant of his guest delivers the message: no gifts for my boss, please, for he will bring you no gifts, also. What is the greatest insult for a guest? It is when his enemy appears in his host's house the same day the guest leaves. So President Rafsanjany of Iran comes one hour *before* the U.S. Secretary of State leaves. Thus, what happened in the city of Almaty was an exchange of insults. The Secretary of the State Department started it unexpectedly; the Kazakh president did it intentionally, but he had no choice. (This is easy to explain: his power is far more important to him than relations with the United States. If he wants to keep his presidency, he must follow the habits of his people.)

Obviously, foreign politics has two goals: first, the smooth running of the routine machinery of international relationships and, second, the invention of new initiatives for the sake and advantage of your country. Toward the newly opened world of Asia there is no old machinery, and by now former Secretary of State J. Baker's initiatives in establishing personal relations with the local leaders have been extinguished.

Is this of any importance for the United States? Let us have a look at this list: Karachaganack (Kazakhstan) — one of the biggest oil-gas-condensate fields in the world — is British and Italian (a $650 million project); Kum-Tor (Kirgizstan) — the third largest gold deposit in the world — is Canadian[3] (a $470 million project); a huge refinery in Uzbekistan (a $1.1 billion project) is Japanese; the promising region around the Aral Sea (a $2500 million project) is French. You can find a continuation of this list in the following chapters. And, of course, I don't know all the projects which exist. While traveling in central Asia, I have met representatives of 56 foreign companies (I checked the business cards given to me), and 39 were Americans (large and small companies). Thus, it is clear that the United States does have interests within this mid-continent.

I know for sure that the Russian, Kazakh, Uzbek, and Kirgiz people are willing to deal with Americans. Of course, they aren't competent within the jungles of Western business. But since the Americans aren't competent within the treacherous steppes of Oriental business, a *Competition in Incompetence* has begun, and at first the different cultures each face the negative side of the other. Official diplomacy should prevent this from happening but has failed.

---

3. The Parliament of Kirgizstan didn't approve this deal.

> "America was a legend among our people. Mr. Baker's visit furthered this idea since he understood the spirit of personal relationship. Nobody exploited this issue further and it is dying . . . together with business, believe me. Well, we'll survive . . ."
>
> (F. Kulov, Vice-President of Kirgizstan; conversation with author, August 1992)

Of course more contact has occurred between these two civilizations since their first business meetings in 1992. But I am certain that Westerners feel a lack of experience is a big obstacle when visiting, negotiating, and establishing business there. What could possibly meet this lack?

The answer is *knowledge:*

- of the historical and political background of these newly opened countries. This would be helpful when involved in unpredictable events (the appearance of which is absolutely predictable);
- of the system which was in use and motion during a long period of time. The local people know only this system and nothing about others;
- of mistakes already made by others (mistakes made by compatriots and by local people);
- of methods of recognizing one's own mistakes as soon as possible and of dealing with the consequences;
- of different opinions on the potential possibilities of conducting business in these countries.

It would be better if all of these features were extracted at another's expense rather than by personal experience.

> "I didn't want to start a diplomatic incident *(by refusing to cut off an ear of a sheep's head and to swallow an eye).* I was mercifully spared *(since he avoided these experiences).*"
>
> (A. Gore, Vice-President of the United States; interview with reporters, Almaty, December 1993)

At least he knew what those features were which he was trying to avoid.

Well, now it is your turn to know these features by reading the next five chapters.

Chapter 1

Kirgizstan: Gold Rush

The Republic of Kirgizstan is located in the southeast corner of the former Soviet Union (FSU) surrounded by Kazakhstan, Uzbekistan, Tajikistan, and China (Fig. 1–2). Situated in the middle of the land mass considered Eurasia, with no ocean coast or plain closer than a thousand miles in any direction, the mountains of Tien Shan (Heavenly Mountains in Chinese) dominate the scenery, the climate, and the life of this country. There are far more sheep than people and more people then arable land in Kirgizstan.

The population of this republic is almost 3.5 million people, about 65% of whom are Kirgiz who believe that they are Sunni Muslims. This religion doesn't have deep roots among these disengaged nomads, and the more influence the fundamentalists gain among the neighboring countries, the more the Kirgiz will resist being swallowed up by such a movement.

Different Kirgiz tribes continue to hold a deeply felt tradition in the sanctity of their earth — namely, the peaks of the mountains, the lakes, trees, and woods and their native animals such as the wolves, bears, and especially the snow leopards — believed to be the ancestors of different Kirgiz tribes.

Less than 40% of the population live in the cities, and less than half of the city dwellers are Kirgiz. In the beginning of this century, Russian villages were established within the arable lands, and even now the Kirgiz minority numbers no more than 30%. Thus, it can be seen that most of the Kirgiz remain engaged in various forms of animal husbandry and related occupations. This explains why the Kirgizian Parliament opposed the privatization of the arable land; most of it would belong to Russian and Uzbek peasants who are skillful in agriculture.

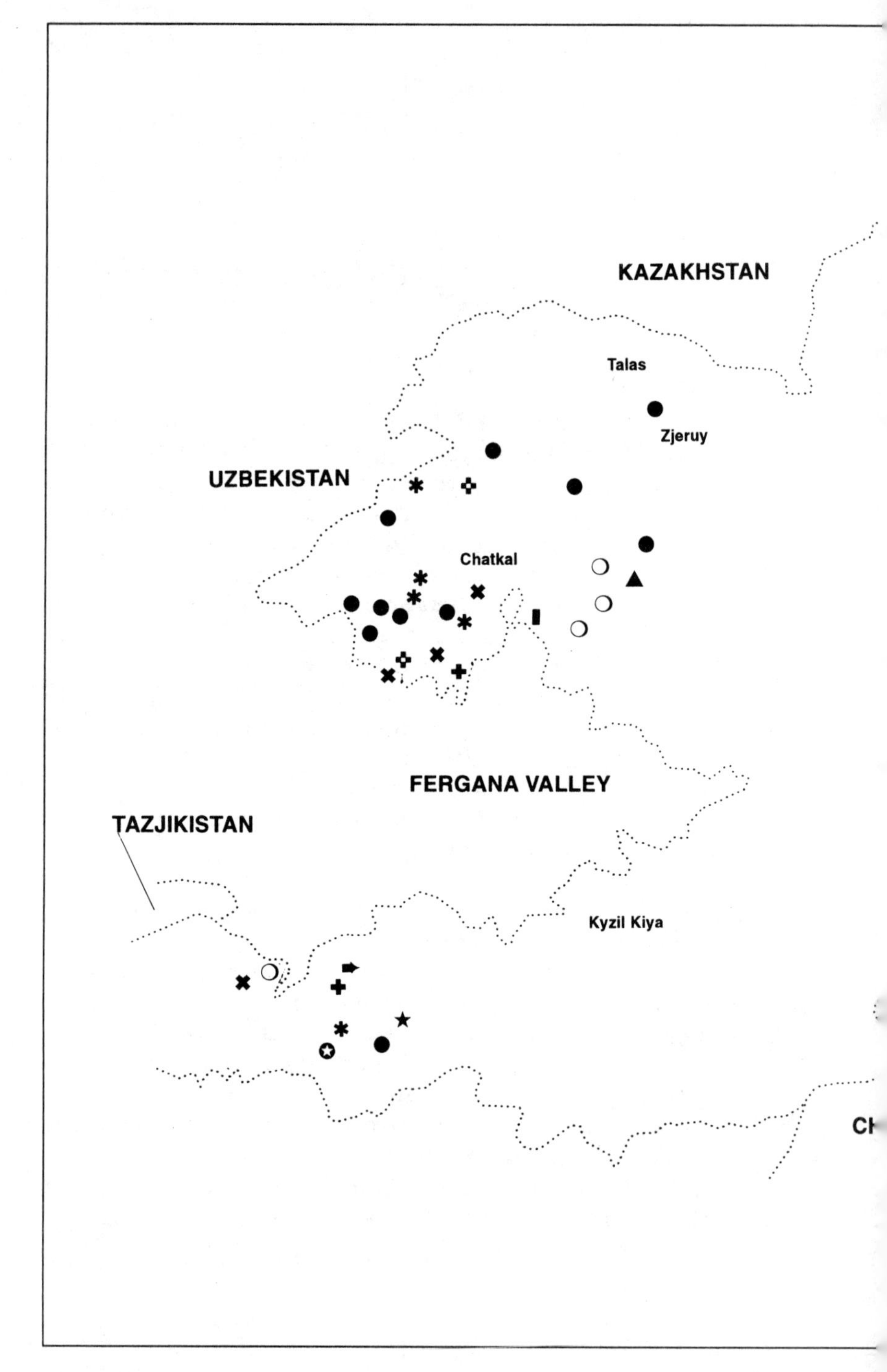

Figure 1–2 Mineral Resources of Kirgizstan

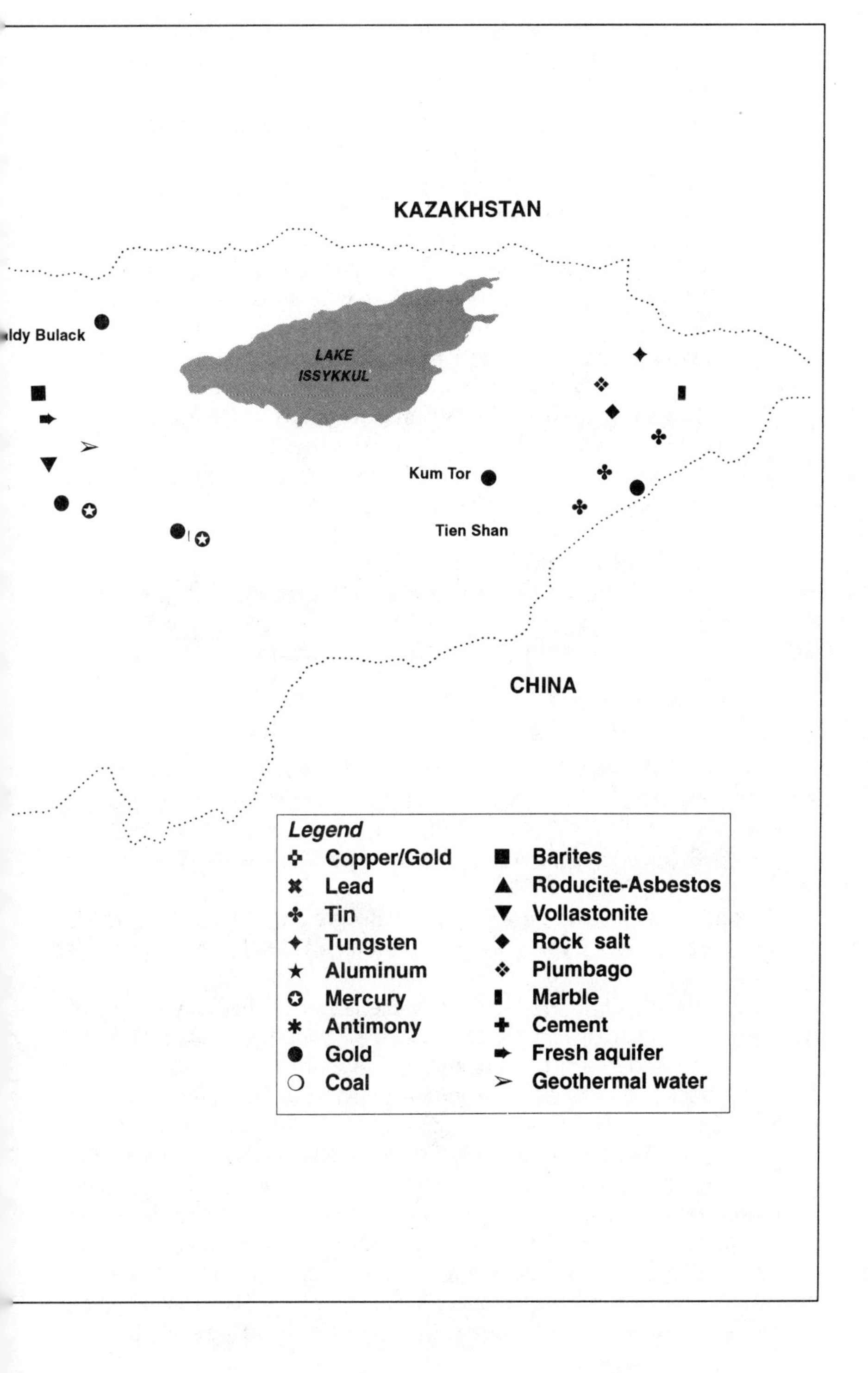

KAZAKHSTAN
ldy Bulack
LAKE
ISSYKKUL
Kum Tor
Tien Shan
CHINA
Legend
Copper/Gold
Lead
Tin
Tungsten
Aluminum
Mercury
Antimony
Gold
Coal
Barites
Roducite-Asbestos
Vollastonite
Rock salt
Plumbago
Marble
Cement
Fresh aquifer
Geothermal water

"We could not privatize the arable land because it would not belong to the Kirgizian population. We have not inherited this problem from the Communist era; it was there long before, when the Russian Empire dispersed the conquered land among the soldiers of its colonial troops. The great-grandfathers of the Russians who live with us now were the first landowners, indeed, because we — Kirgiz — were nomads in our country. These Russians are citizens of this country, and it will be better if we keep the collective farms, where these Russians work together with the Kirgiz. There is no other way to avoid a civil war. But nothing else can be privatized if the land is not. So we are doomed to keep an inefficient economy and to live together in poverty. If you Americans understand the matter, come and help us. I am afraid you will not."

(Bolot T. Shamshiev, famous movie director, former member of the first and the last Soviet Parliament, member of the Kirgiz Parliament; interview with reporters P. Fedynsky and N. Sorokin, Voice of America, June 1992)

◆

Between 1983 and 1988, less then 300 foreigners (Czechs, Germans, Americans, French) visited Kirgizstan. In 1989 there were more then 500 foreign visitors; in 1990, about 2,000; in 1991, more than 5,500; in 1992, more than 25,000 (Japanese, Germans, Turks, Chinese, Americans, Canadians, British, etc.).

Among the visitors during 1983 to 1988 were 3 officials and 3 businessmen; in 1992 about 200 officials and more than a 1,000 businessmen visited Kirgizstan.

Obviously, the Kirgiz are very hospitable like any other Oriental people. However, you should have iron nerves while enjoying their hospitality.

These descendants of nomads revere their domestic animals as friends. They must persuade this sheep (which after one hour will be eaten by the hosts and guests) that it is sacrificed to the guests, who — the host believes — are sacred. To persuade the meek animal, the host requires the presence of the guests when he cuts the throat of this sheep. You may avoid this show simply by asking for water or tea or kumys (sour horse milk) since the guest's requirements are sacred also. What is unavoidable, is the presence of the sheep's head on the table watching whether the guests really do enjoy eating its body. Then one may have the sheep's eye in his hand suddenly, or an ear. They expect the guest to give a short speech and then hand the ear to a chosen friend. The speech explains why the chosen person deserved this strange but sacred gift. After this gift is eaten, the soul of the killed creature is satisfied, and the friendship between man and animal is reestablished.

None of this includes women, though they may be present nowadays. As a guest, she is sacred, but she would be more respected the more she pretended to be invisible. It is a complicated game but, if successful, she will gain a special fame forever and everywhere. An irrevocable support of *ak-sakals* (white-bearded seniors) would escort her anywhere, including the President's Palace.

For example, a small company (six employees) located in Singapore earned a few million dollars during the year 1992 trading Russian aluminum through Kirgizstan and China because their female translator (a Chinese girl) behaved properly while a guest (among others) in a *yurt* (Kirgiz nomad's tent).

The governor of a province, in league with a British businessman, was a competitor of this small company. The governor knew exactly who in the government had to be bribed. He spent almost a week with this Singapore delegation; he drank a lot of alcohol with them; he had eaten almost 10 sheep together with them in different yurts. He tried to keep the men from Singapore out of Bishkek, the capital of the country, until the officials there would sign the documents necessary for him and his British companion. In vain!

The female translator always acted properly in those yurts, and the *aksakals* liked her, and she had the deal.

Soon this governor was supposedly promoted to be a minister in the government. The position carried with it no responsibility and, recently, the parliament eliminated both the ministry and the position.

◆

> "Oh, our respectable President Akaev does not like even a smell of our autonomy. We have oil in our region. We have gold in the Chatkal area. We want this for our people. I negotiated with some solid foreign representatives these deals. He himself rejected those. Does he want to appoint the partners for us by himself? I was in the USA twice and he was not. Does he know these companies better? No, he does not. What is going on here? He hates the governors trying to make our people rich, he hates us making decision by ourselves. Isn't this wrong?"
>
> (B. Osmonov, former Governor of the Zjalal-Abad region, member of the Parliament of Kirgizstan; speech to Parliament, September 1992)

"I like rich regions; I like rich people. What I dislike are rich governors. This country must be united."

(Askar Akaev, President of the Republic of Kirgizstan; conversation with author, June 1992)

◆

Kirgizstan is rich in mineral resources (Fig. 1–2). Local geologists have made a very reliable exploration within these elevated plateaus, valleys, and slopes—in some areas covered by snow almost nine months of a year.

A few deposits of copper, lead, tin, tungsten, aluminum, mercury, antimony, gold, etc., are found in Kirgizstan. More than 50% of the uranium ore produced in the FSU was obtained from a mine near the city of Kyzyl-Kia (Fig.1–2) in southeastern Kirgizstan (Osh region).

In geological terms, the territory of this republic presents a folded region composed of geosynclinal and orogenic formations of the Mesozoic and Cenozoic age. All of Kirgizstan (as a part of the FSU) was mapped and covered by geological survey in the scale of 1:200,000 (which means that the distance between geological traverses isn't more than 2.0 km, and the interval between observation points isn't more than 0.5 km). Some large promising areas are mapped in the scale of 1:100,000 (the profile spacing is 1.0 km, and the interval between observation points is 0.2 km). At the deposits, geological survey and measurement were made along rectangular profiles. The profile spacing is 0.1 km, and the interval between measurement points is 0.025 km. Thus, the survey was performed according to the strict standards of the FSU, and the data on proved reserves of the ores and minerals are reliable.

The commercial reserves of gold in Kirgizstan are about 1,200 metric tons.

The Kum-Tor deposit (Fig. 1–2) contains as much as 850 metric tons of gold with a concentration of more than 11 grams of gold per ton of ore. As multiple veins occur deeper under the surface, the more gold the ores contain. The Kum-Tor gold veins are dispersed as deep as 450 meters in a spot as long as 950 meters and as wide as 300 meters (Fig. 1–3). The Kum-Tor gold deposit is elevated as high as 4,550 meters and is located in an uninhabited upland with a rigorous climate. The deposit has been in exploration during the years 1982 to 1989, and no development was started due to political events in the FSU. The Kirgiz government has neither the money nor the specialists and proper equipment for developing this deposit.

The Taldy-Bulak deposit is located within a lovely valley with a moderate climate (Fig. 1–2). The deposit contains more than 100 metric tons of gold with a concentration of more than 9 grams of gold per ton of ore. Again, the gold ores deeper under the surface are more rich. Something like 10% of the gold ores were produced out of this deposit using an open pit method during the years 1989 to 1992. This is the only gold deposit which actually produced gold in Kirgizstan. The figures of production are secret, of course, but in December 1992 President Akaev told parliament that the government had deposited 2.5 metric tons of gold in a bank in Switzerland.

Figure 1–3 Geological Layout of the Central Area of the Kum-Tor Gold Deposit

The Zjeruy gold deposit (Fig. 1–2) contains more than 80 metric tons of gold with a concentration of more than 16 grams per ton of ore. It also is located within an upland whose elevation is less than 1,600 meters. This deposit has been explored through the years 1986–1990 and was never put into production.

The rest of more than 100 metric tons of the gold stock in Kirgizstan is dispersed among 11 explored deposits in the Chatkal region (Fig. 1–2). The estimated reserves there might be 10 times more.

If Moscow's *Soyuzalmazzoloto* (Union's Diamonds and Gold) enterprise were to exist and operate during the 1990s as it had before, these three large gold deposits would have been developed until the year 1996 and the smaller ones until the year 2000. These were the plans.

Moscow's headquarters would provide the money, the specialists, and all the equipment and machinery.

The general director of each local developing enterprise would be a Kirgiz (a member of the Communist Party) and the personnel (miners, engineers, clerks) would be mostly Russians and some people of other FSU nationalities. The chief of the miner's party organization would also be a Kirgiz as would be the mayor of the newly established small city and his personnel (all of them members of the Communist Party). These two individuals would be responsible for the assignment of housing (by law, a miner's housing would be free of charge), and these were the positions especially predestined for bribes because of the everlasting deficit of homes and apartments. The supply chief and personnel, responsible for food and goods, would be Kirgiz also. There is a clear possibility for one in this position to steal the food and the goods coming from abroad (and also to give them to chosen persons such as the general director or the mayor, etc.). The privilege to appoint individuals to these positions would be given to the authorities of the republic. Whoever wished to be appointed would have to pay for the privilege — money and especially goods (from Japan for the family and from France for wives and daughters) would be preferable. To be in command of the gold resources, the bosses in Moscow would wish to have a lot of happy people here. They called it "clever national policy of our party."

And it was clever enough. Nobody in the republic, including the prime minister and the first secretary of the central committee of the Communist Party of Kirgizstan, would know how much gold had been produced here and where it had gone. But in exchange for this gold, Kirgizstan would obtain from the Moscow authorities supplies of bread; gasoline, diesel, and jet fuel; aircraft and helicopters; TV sets, personal computers, phones, faxes, and other goods the country badly needed.

After the dissolution of the FSU, the new Kirgizstan government did not bother with development of their gold deposits. First, the bureaucrats within and around the government were busy struggling for power, and second, the

Kirgizs had no first-hand experience in such development and lacked the know-how in international relations, trade, and business to obtain foreign partners.

Those were some bonafide international adventurers who first came to try their luck.

◆

> "Anybody who has a few dollars to spend on airfare there and back should come to Bishkek to look for his chance."
>
> (President of a small American oil company; conversation with the first U.S. Ambassador to Kirgizstan, August 1992)

An incomplete list of foreign companies who sent delegations to Bishkek during the year 1992 includes: the Siabeco Group; HEMCO—Hunt Exploration and Mining Co.; Brush Creek Mining and Development Co.; Cameco (TSE); Minproc and Chilewich; Amax Exploration, Inc.; Newmont; ASARCO; etc.

The government was rather perplexed. A short excerpt from the author's (A.) conversation (August 1992) with Vice-President Felix Kulov (F.K.) of Kirgizstan demonstrates their perplexity:

F.K. Too many of them, too much. It is a new problem: how to choose with no information on them? And they are spoiling our people — gifts, bribes . . .

A. It was always this way before, wasn't it? . . . Yes, you will probably make mistakes whichever company you choose. And this is your payment for the experience.

F.K. Spoiled and experienced people. It is even worse, isn't it?

A. No, it isn't. The alternative is to forget the gold, leave it where it is now . . .

F.K. . . . and the poverty of this country. Do we understand this dilemma? Yes, we do. We will survive. . . . So I want to tell you, we will be a long time waiting.

A. Well, it is your way, instead of organizing a civilized tender. Will you wait to the limits of their patience? They say in America: time is money.

F.K. Oh, yes! We have neither time nor money. I want the companies to inform us about each other. We can wait.

And the companies do, and the Government waits.

◆

The area along the northern, eastern, and southern edges of the Fergana intermountain basin belongs to Kirgizstan (Fig. 1–2). It is the only place where oil is produced in this republic.

The depresson of the central Fergana basin is as deep as 8.6 km, and the Paleogenic and Neogenic terrigenous sequences (mostly continental sedimentary systems) fill it out. But towards the edges, the depression becomes more and more shallow, and those deposits pinch out and rise to the surface. Because of the various tectonic movements, the geological structure of the eastern part of the Fergana basin is extremely complicated and still unclear. At the Kirgiz portion of the Fergana basin, the last exploratory well was drilled in 1979 and the last developing well in 1987. From the point of view of the Moscow-based Ministry of Oil of the FSU, it wasn't worthwhile investing a lot of money in the exploration and development of oil fields within the Fergana basin because the estimated reserves were comparatively small (less then 1 billion metric tons in-place).

From Moscow's point of view, it was more reasonable to invest up to 85% of the ministry budget in the development of the huge West Siberian basin. But in the beginning of the 1980s, the Chinese discovered large oil fields in the area close to the border with Kirgizstan (no connection to the Fergana basin). It was a political challenge, and the Moscow authorities were very sensitive to such issues and found the money for Kirgizstan. A geophysical company, *Saratovneftegeofizika*, located in the city of Saratov on the Volga River, brought their personnel and equipment to Kirgizstan, organized a base, and began the seismic survey. As many as 42 seismic lines (590 km) across the Alay depression, 37 lines (260 km) across the Chu depression, and 880 lines (7,229 km) across the Kirgizstan part of the Fergana depression were worked out, covering an area of more than 11,400 square kilometers. The drilling of a deep (deeper than 5 km) wildcat started in the Alay depression, and Saratov's geophysicists were responsible for well logging there also. Some copies of short reports were sent to a local oil company in Kirgizstan (which, like the geophysicists, also reported to Moscow), but nobody in the oil company or in the government took the effort to pay attention to this information. They had nothing to do with it at all and, indeed, since all decisions had been made in Moscow, somehow plenty of fuel was delivered to the republic.

During the years 1980 to 1990, Kirgizstan produced less then 1.9 million metric tons of oil; during 1991, less then 125 thousand tons; and during 1992, approximately 87 thousand tons, all from more than 60 wells. No one can give more accurate figures because any calculations made by the local specialists should be questioned. Misrepresentation of statistical reports was a very old FSU political and economic game; the salaries, positions, and decorations of the local leaders depended on the written figures, not on the real production. Finally, nobody knew what was going on in the industry.

Oil from the Kirgiz oil fields is delivered to the Fergana refinery located in Uzbekistan, 100 miles to the west, and the products (gasoline, diesel, jet fuel, etc.) are brought back to Kirgizstan. General directors of both enterprises (the Uzbek refinery and Kirgiz oil company) were Armenians, and, of course, they were close friends. So they managed to have their statistical figures in accordance and to supply Kirgizstan enough gasoline. Nobody knew the quantity of the additional gasoline, which indeed had been produced from Uzbek oil and then was sold in Kirgizstan.

At the end of 1991, the republics of the FSU separated, and both general directors were replaced. (Both of them left central Asia.) An Uzbek became the chief of the Fergana refinery, and a Kirgiz became the chief of the Kirgiz oil company. In early 1992, the government of Kirgizstan was surprised to find the country facing a 65% shortage of gasoline supplies which reached 80% by the summer.

The presidents of the two countries, Mr. A. Akaev of Kirgizstan and Mr. N. Nazarbaev of Kazakhstan, are friends also. They met twice during the first half of 1992, and Kazakhstan agreed to supply some petroleum products to Kirgizstan.

But Kazakhstan's richness didn't help Kirgizstan. During the first half of 1993, only 25% of the buses, 33% of the taxis, 20% of trucks, and less than 50% of the tractors were in operation in Kirgizstan. In 1991, flights from Bishkek to Moscow were scheduled twice a day — in 1993, once a week; to Tashkent in 1991, three times a day — in 1993, once a week; to Istanbul in 1992, twice a week — in 1993, none; to China in 1992, three times a week — in 1993, none.

The price of gasoline in 1993 became 1,500 times higher than in 1991, and the price of diesel was 10,000 times higher.

Thus, the friendship of two presidents doesn't appear to be as effective as the friendship of two general directors.

◆

"Here we are, and I have discussed this matter in the governmental circles. We are ready to give you 49% of the increased production of the oil, if any. Come and do the work in our oilfields. What will be your profit? How do I know?"

(T. Kholbaev, General Director of the state oil company *Kirgizneft*; conversation with author and a vice-president of a small American gas company, May 1992. Two days before, Mr. Kholbaev had returned from New York where he had attended a one-month business course at Columbia University sponsored by the World Bank.)

◆

It is time to have a quick look at Kazakhstan, the country which is second in area to Russia but whose population (16.5 million people) is almost 10 times less than Russia's. The native population — Kazakhs — are Moslems of Mongol descent but speak a Turkic language which is almost the same which the Kirgizs speak. Due to the "wise national policy" of the Communist Party, Kazakhs form a minority of 48% in their own country.

Under President Nazarbaev, Kazakhstan has become the leader in attracting foreign investments.

◆

> **"Still, the Kazakh Government has provided the crucial ingredient, political stability."**
>
> (M. Dupre, General Director, *Tengizchevroil*; interview with a U.S. newspaper, March 20, 1994)

In contrast to other regions of the FSU, Kazakhstan's oil production increased from 2 million tons in 1965 to 22.8 million tons in 1985 and to 26.6 million tons in 1991. The development of Kazakhstan's oil fields, located in the Pre-Caspian Depression and at Mangyshlack Peninsula, began only recently, and less than 10% of the recoverable reserves have been produced.

Chevron has set a model of success in developing the Tengiz oil field (northeast coast of the Caspian Sea) which is estimated to be about the size of Prudhoe Bay in Alaska. The oil contains sulfurous contaminants that must be transported in Russian pipelines. Although Russian oil is highly sulfurous also, the Russians decided that they don't want Chevron's sulfurous oil in their pipelines. Chevron started the construction of a special plant to take out most of the contaminants even though the company feels that it is receiving lower-quality oil in the Black Sea port of Novorossisk than was put into the pipeline network.

Russia's political games are crystal clear: Kazakhstan shall not feel too independent. Still, Russia's will is a reality which pays to be known. And, once again, it has nothing to do with economical problems or reasons.

All across the FSU you will meet this stereotype of thinking: policy predominates the economy.

◆

At the end of the 1980s, political changes were accelerating, as happens when a great empire collapses. The large and sluggish Western corporations were groping unsuccessfully to find the proper strategy for their activity within the newly opened continent. But it was time for tactics — for simple ideas and uncomplicated schemes. Perhaps, this time will last until the end of the century.

Since it became independent in 1991, Kirgizstan (as well as the Baltic Republics) immediately became a staging post for raw materials on the way from Russia abroad through China. It was easy to obtain permissions by bribing the officials, and very modest custom fees were imposed, if any. But in contrast to the Baltic Republics of the FSU, no oil trade was involved in Kirgizstan because it had no pipelines.

Some excerpts from an interview, given by the author (A.) and the Vice-President (V.P.) of a small American gas company (this was his first time in the FSU) to a reporter (R.) for the *Molodezj Kirgizstana* (May 1992, printed June 1992) follows:

◆

A.: Yes, you have faced a situation where many adventurers, ignorant merchants, and aggressive small companies would be involved. Just accept it. . . . Try to use them to pull in the Western elephants with real money. Then . . .

R.: I cannot believe we have the crocodiles here now, have we?

A.: And them, too. You will have to choose those who aren't hungry and make them share with you the mutual success. Most of them are willing to act exactly this way.

R.: Here we are very suspicious on the issue of colonialism: robbery of our national property, environmental problems.

V.P.: Not these. . . . The well at the Minbulack area in the Uzbek portion of the Fergana Valley exploded and wasted up to 150,000 barrels of oil a day during the whole month of March. This is a disaster. If we work here — no matter what the conditions are — we would not spill a drop on the ground. The equipment you use here is not just old, it's ancient. It takes you four years to drill a deep well; we do it in a few months. And take a look at the Saudi Arabians, Kuwait. They are rich, aren't they? Of course they aren't located in the middle of nowhere like you are.

R.: Is it a problem?

V.P.: Yes, it is. We will share the increased oil production, but where shall we sell this oil, since your money isn't real?

R.: I thought it is.

A.: Not for Americans.

V.P.: So we want to put the oil on the international markets, but there is no way for delivering it there.

R.: Is this why you have brought with you these complex plans on oil and gold properties and pipeline construction?

A.: My plan was, obviously, large. As a matter of fact, I think it was too large.

R.: Is there a chance it would start soon?
A.: Not very soon, I assume.

"Do I want to invest money in Kirgizstan? No, I don't. But suppose that I obtain some partners ready and willing to go there for the oil or for the gold. . . . By the way, nobody knows how much oil and how much gold they have there. If I find such partners, I would spend some money for the pre-feasibility study. Yes, I would . . . and we will see then.
What I really want is the exclusive rights on the oil, and especially on the gold. It doesn't matter that this means nothing in Kirgizstan or elsewhere within the FSU. It does matter here. If I have a document proving these rights, I can wait until somebody finds Kirgizstan feasible. Then this document becomes my money. Or the Court will make it so."

(President of a small American gas company; conversation with author, March 1992)

◆

The following excerpts are from a meeting with Mr. T. Kholbaev (T.K.), General Director of the Kirgiz state oil company *Kirgizneft*. The American representatives were the vice-president (V.P.) of a small American gas company and the author (A.). (May 1992, city of Bishkek, Kirgizstan)

◆

1.

T.K.: I cannot understand why you keep insisting on the *exclusive rights* to make the exploration and production within all our petroleum basins.

V.P.: Because we wouldn't like it if another company came to negotiate this matter and you changed your mind.

T.K.: It is true that if someone brings a better proposal — it would happen. We discussed a preliminary document, didn't we? Why should I limit my options for the future?

V.P.: Under this protocol, we bear important obligations regarding the collection of geological information, the reprocessing of seismic data, payment of your delegations' expenses while in the United States, and so on. It costs real money, doesn't it? Preliminary money doesn't exist as the preliminary protocols do.

T.K.: We have already spent real money to produce the information, seismic survey, and so on. We will discuss this matter later. Back to exclusive rights: I want to lose the word *exclusive* in here.

V.P.: No way. It is the most important issue for my company.

Maybe it is true, but why not use the words "very important," which is true also? For Mr. Kholbaev, "most important" sounds a little strange in this context because for him, something other than the "exclusive rights" is the real importance in this deal.
Also, it is almost impossible for these ambitious Orientals to continue a conversation when the word no *has been pronounced, and it doesn't matter who said it first — the meeting is over.*

A.: Mr. Kholbaev, can we have some more tea, please?

An absolutely essential ploy —for a guest to ask an Oriental host for something which he can provide as a powerful symbol of hospitality — to be used at a critical juncture, especially when a blunder is about to derail the negotiations.

That night I invited my friend Mr. Bolot Shamshiev to join us for dinner with Mr. Kholbaev at our hotel. Mr. Shamshiev is a famous movie director on the European scale and, together with a prominent writer, Chingiz Aytmatov, became a national symbol for the Kirgiz people. After dissolution of the FSU, Mr. Shamshiev started a successful trade business trying to collect money for his future movies. I have explained to Mr. Shamshiev the problem with *exclusive rights* and he understood that it means nothing to the Kirgiz since it touches the matter of entire competition between foreign companies. Mr. Shamshiev and Mr. Kholbaev had a short, private conversation after the fourth toast to mutual cooperation. The problem of *exclusive rights* was solved.

Rule 1–1

Establish a personal relationship with local people while you have no problems, because that makes this relationship reliable; use personal relationship when you face problems; it strengthens this relationship.

Within the Western culture one is a partner and then may become a friend. Within the Oriental culture one is a friend and then may become a partner.

2

T.K.: Yes, we have the figures of the reserves of the oil in-place and recoverable. These are secret and I cannot disclose or sell them, of course. But . . .

VP.: My company wouldn't buy it. These data are the keystone of the entire deal, and I want them for free.

The interrupted general director wanted to say: But if I have permission from my government, I will, probably, sell you all the data.
In central Asia it is not impolite to interrupt another's idea. The local people use this ploy often as a sign that they are willing to continue the conversation. Mr. Kholbaev didn't believe that the V.P. was being truthful and, contrary to what he said, those figures were already known by him.
As was mentioned above, the estimation of the oil reserves in Kirgizstan had been rather modest. After an explosion[1] of a deep well in the Minbulack[2] area (Fergana Valley, Uzbekistan), the reserves estimation tripled. The interrupted general director was assured at this moment that the American specialists had knowledge of some large figures, and they came to Kirgizstan because they believed that the oil resources are huge. "Why does this V.P. hide that he was informed?" From now on the V.P. will not be trusted any longer, and his presence will be harmful in any negotiations with Mr. Kholbaev. It was not the essence of this deal that affected Mr. Kholbaev unfavorably but the displayed reticence *of the V.P.*

A.: Mr. Kholbaev, I am sure you have some reports on the reserves' estimation from different Moscow institutions. As I understand, they are as secret as the final figures. But the data on porosity, oil-saturation, gross pay thickness, area, etc., aren't secret, are they?

T.K.: No, some of them aren't if given separately.

A.: This company is willing to count the reserves themselves. How about that?

T.K.: I don't have a problem here. I will have a girl write down some of the figures for him.

---

1. The event was reported by the mass media all over the world.
2. Less than 30 km to Kirgizstan's boundary.

You should become familiar with the so-called system of secrecy which was established in the FSU.

> "We have paid a lot of money for their data on Azeri (an offshore oil field within the Caspian Sea) and we revised them. First, they were incomplete; second, they were wrong; third, they were on another field."
>
> (Vice-president of a large American and international oil company; conversation with author after a trip to Azerbaijan, April 1990).

VP.: That secrecy was in circulation under the old FSU laws. Why should you follow it now?
T.K.: Nobody has revoked these rules.
V.P.: Since you are now an independent country, you do it.
T.K.: It is not my business, certainly.
A.: Right you are. You wouldn't object if we discuss this problem with your prime minister, would you?
T.K.: Not at all, of course I would not.

It is useful to mention a few times during the negotiations that you will meet (or have already met) a very important person. Be careful if you know that your interlocutor was appointed to his position by another VIP. These different VIPs and their teams (be assured) are enemies, so one clan would like to deny the results of your meetings with the members of another clan in spite of the harm done to the business. These people have their own businesses, which for them are far more important.

A.: You know, I mentioned this problem when I called your Minister of Industry, Mr.Omuraliev. As I understood, he is not the one to solve it.
T.K.: Him? I am under him for a few weeks. Soon we will have the Energy and Fuel Ministry.

Being a man of the prime minister he saw himself in this position, of course.
As a matter of fact, I was told by Mr. Omuraliev (his formal boss) that Mr. Kholbaev is able to arrange by himself any data he has. I didn't mention that because Mr. Omuraliev was a man of the vice president — so the two gentlemen belong to different clans (and tribes).

Rule 1-2

Use personal relationships for searching into the background of your intended interlocutors to determine the identity of his real (not formal) boss. Appeal to this boss when having problems in the negotiations.

The Oriental hierarchy is far more complicated than the Western, where most often the formal and real boss is the same person. Be aware that in central Asia the reports sent to one's formal and real boss are quite different — as their orders would be different.

V.P.: Well, you have promised me the well-logging curves, and I saw them. They contain no useful information. Do you have something more current?

T.K.: Yes, we do. They are in the city of Saratov in Russia. They have, as you know, the seismic information also.

V.P.: Can we have those?

T.K.: Yes, of course. We will try to have all these data here as soon as possible and you can buy whatever you will choose. I have sent a letter to Saratov already.

VP.: They claim in Saratov that this information belongs to them and that they are the owners. Are they? They want to sell it themselves.

T.K.: Don't buy it from them. We would not appreciate this deal, believe me. We took the base they had here; it is ours now. And the information is ours.

> "Theirs! They didn't pay a penny either for seismic exploration or for well logging. It was money from the Union Oil Ministry. . . . and they claim our geophysical base there as being theirs in spite of my company's having invested money there a few years ago. Suppose their president writes a letter to Mr. Yeltsin and Yeltsin himself orders me to send the data to Kirgizstan. Will I? Take a guess? No way."
>
> (A. Michurin, President of the Joint-Stock Geophysical Company *Saratovneftegeofizika;* conversation with author, June 1992)

V.P.: For God's sake, where is justice? We came to invest money here, and we have to pay for the data. Are you interested in our investments, Mr. Kholbaev?

He personally is not. Nothing there — including the oil fields, oil, land, and the rigs — is or will be his private property. Since the professionals (who aren't Kirgiz) have left the site, he and his subordinates have no idea how to operate the rigs or produce the oil. So he has a goal (not an interest) to remain a commander of a brigade which works at the oil fields.

T.K.: I am here talking with you; so I am interested in some of your proposals for God's sake.

A.: Justice and injustice are human's; God's sake is above. . . . Let us discuss justice. As you can see, Mr. Kholbaev, we have three choices. The first is: we obtain the data from *Saratovneftegeofizika* and start working; second, we wait until you receive the data, but, as I believe, it won't happen very soon; third, we try to help you in Saratov, and then you will give us the information for free.

V.P.: The protocol states that you have to give us the information for free. I have no intention to discuss it here.

Why not? The protocol isn't signed yet, and each sentence is a subject for discussion.
Another remark: You should work as a member of a team. The hosts will not respect (which means they will not take seriously) several people with different opinions, and your company's reputation will be ruined for a long time. If it happens, leave this country.

Rule 1-3
At the negotiating table work as a team.

At the table, discuss any problems among you openly and loudly, and the translator may explain to the hosts the matter of your discussion. But no contradiction is acceptable in the statements of the different members of the delegation to your hosts.

T.K.: I didn't sign this protocol.

Don't expect him to say: I haven't signed this document yet. The starting position of these people is very natural. To them the position is: "You came up to me; you need me but I am not sure I need you here and your businesses. Nevertheless, since you are my guests, try to use this advantage." PrimaOnce

rily, this position has nothing to do with the business. The process of Oriental negotiations pretends to be an action during which the guest tries to involve the host in the sphere of his (guest's) problems.
Very important: When you change places and he is your guest in the United States, his tactics will change and he will try to win the initiative in the negotiations.

Rule 1-4
Invite your partners to visit your country, and organize the visit as soon as possible.

In your country, you will be amazed at meeting such changed persons now willing to discuss business in the best way.

A.: Mr. Vice-President, let us insert a neutral word here: neither the word *sell* nor the words *give for free* should be inserted, but the words *shall provide* are appropriate. First we shall speak with Saratov's geophysicists. How does this sit with you, Mr. Kholbaev?

Immediately discuss a problem at the table and involve the host in the solution. This shows him that he is equally responsible for the success of the negotiations.

u

Before rumors of the great gold reserves in Kirgizstan had reached other countries, and the week after President Akaev declared the independence of his country, the chief of a Swiss banking group came to see him. The former President of the Science Academy of Kirgizstan (A. Akaev) and the former emigrant from Lithuania to Israel, then to Canada, then to Switzerland (the Swiss banker) soon became close.

His enemies say that this Swiss banker is the chief banker for the Communist Party's money and a colonel of the KGB, though nobody could prove it. The facts are that the banker was and is a close adviser to: President A. Akaev of Kirgizstan, President M. Snegur of Moldova, President A. Brazauskas of Lithuania, Vice-President (former) A. Rutskoy of Russia, Vice-Premier Minister (former) V. Shumeyko of Russia. These persons above have a lot of enemies and so does the Swiss banker.

It is a common ploy in the described region to talk about scandals. Just keep in mind the above list of men while trying to establish businesses in those countries.

The Swiss banking group helped President Akaev to advertise Kirgizstan's richness abroad, to establish a Kirgiz currency, to transfer state gold reserves into a Swiss bank, and to visit Israel. As if this were not enough, they helped Mr. Akaev's daughter become a student in a Swiss University.

Newmont, a world-renowned mining company, came to Kirgizstan after the Swiss banking group and did not succeed in any gold deal.

◆

Be prepared to face the fact that competition is already fast and furious in these newly opened countries.

When rumors of the solid gold resources in Kirgizstan reached different corners of the world, reaction was fast and many foreigners flooded Bishkek.

A mining engineer was hired by a small American gas company; was appointed its vice-president; and sent to the CIS with the president of an affiliated company.

They didn't know whether or not they would need visas (even citizens in the FSU did not know), but to be on the safe side they applied. They were told by the consulate that no visas were required, but then they could not buy airline tickets to Bishkek. A map revealed that Tashkent, capital of Uzbekistan, was nearby. They flew to Tashkent instead of Bishkek only to find that they now needed visas to be in Tashkent. So, they had to return to Moscow. Next day, one ticket to Bishkek was obtained, and the vice-president (mining) flew five hours in an overloaded aircraft. Then he spent a night inside a stuffy airport sitting among his six suitcases; meanwhile, the president of the mining company tried to call from Moscow to warn his colleagues that this vice-president might be in Bishkek. Early in the morning the phone call finally got through.

◆

> "I enjoy traveling in Russia. It doesn't matter whether you are in a bus, a train, a plane, or subway; you are close to heaven."
>
> (P. Buckley, Vice-President, Petron, Inc.; conversation with author, March 1993).

Don't make any appointments until you are actually in the city in which they are to take place.

◆

A meeting was held on June 2, 1992 in the city of Bishkek, Kirgizstan, with Mr. S. Tekenov (S.T.), General Director of the Kirgizian State Committee of Geological Explorations — *Kirggeology*. The American representatives were the vice-president of Exploration (V.P.E.) of a small American gas company, the vice-president of Mining (V.P.M.) of the same company, and the author (A.). This was the Americans' *first* time in the FSU.

◆

S.T.: Have you noticed, gentlemen, how things have changed in our country? A year ago, to meet you here in my office I would have needed first: a telex from the Moscow Ministry of Geology; second, permission from our Republican KGB office; third, a letter from *Intourist* stating that you are staying in their special hotel; and fourth, the police would have had to be informed about the time of your arrival and departure from my office. How would you like it?

V.P.E.: We wouldn't, at all.

He forgot to mention that the telex would read: Don't make any promises; don't sign any papers; and especially make them feel that they have nothing to do here. Also, immediately after the meeting, he would need to write two reports — one for the KGB and the second for the ministry, describing all the issues discussed.

S.T.: . . . Now, are we ready to return to business?

V.P.M.: Oh, yes we are.

He began a political conversation, and even though he asked if you were ready to return to business, it was done out of politeness; in reality, he wanted to know your political opinion.

Rule 1–5

Never initiate discussion on political themes. Discuss politics as long as your host wishes to do so and even longer, if your host started it.

Political issues are extremely sensitive for any manager you meet in the FSU. And within central Asia, it is still dangerous to discuss these issues with foreigners. However, if the host starts the discussion, go ahead, even if you are in a hurry. Remember, for him it is more important than the current business. If he discusses politics with you, it means that he has already approved your deal.

S.T.: . . . So you know that we made all the explorations at the Kum-Tor, Zjeruy, and Chatkal areas. Then, a year ago, we gave the first two to the mining company *Kirgizzoloto* (Kirgizgold), and these belong to them now. We cannot discuss them with you. Are you interested in the Chatkal gold fields?

V.P.M: We certainly are, but we know nothing about those fields.

He should say: We don't know a lot *about these deposits.*

V.P.M.: Let me ask this question, Mr. Tekenov. Have you sold Kum-Tor and Zjeruy, or did you give them away for nothing?

This is a very good question which serves to establish a personal relationship which, once more, is the key issue within this part of the world. The vice-president's question indicated an interest in Mr.Tekenov's situation, his American understanding of the property problems, and his ignorance about the FSU's system of managing the industry.

Rule 1-6
Show your willingness to be a pupil.

It is against your ego, of course. But in the Oriental cultures, the relationship between the teacher and a pupil is most respectable, and your teacher becomes your defender everywhere, also. Your time to teach will come soon, immediately after you have established a business here.

S.T.: We were paid for our jobs by the Soviet government. Now the producing company will be paid by the Kirgiz government. If the exploration had been completed this year, we would never have given Kum-Tor into somone else's hand. But this happened a year ago. . . . Well, the Chatkal is still ours, and we will appreciate your cooperation there. Visit it. My geologists will accompany you. My specialists will give you all the information on Chatkal you need after I have spoken with our vice-president, Mr. Kulov.

He indicated his boss (by name), which means that the boss was in favor of this deal before we had met Mr. Tekenov. If you can read these signs, you will understand what is and is not negotiable and what the potential price range of the deal is. Because he mentioned his boss first, feel free to ask questions regarding him.

Rule 1-7
Discuss the political leaders of the country only after your host has mentioned them first. Discuss only the leader who was mentioned.

In the Oriental cultures the Ruler was, is, and will be sacred forever. One can hate him, kill him, and still worship him as a divine creature.

A.: Mr. Tekenov, as I understand it, the secrecy of the information was established in Moscow. You know that we are interested in oil properties but also in gold. We need the data on Kum-Tor, Taldy-Bulak, and Zjeruy which, I believe, you have. I know that Mr. Kulov was the Minister of Internal Affairs of your republic and now, being the vice-president, is responsible for the issues of security. Is he the person who approves the release of geological information?

S.T.: Well, in my office . . . yes, he is.

"Ores with arsenic in Chatkal; Kum-Tor's elevation. If I told you that I had more information before coming here than now before leaving, it would not be the truth. Almost."

(Nelson B. Hunt, President of HEMCO; interview with a reporter of the *Slovo Kirgizstana*, August 1992)

Mr. Hunt was brought to Kirgizstan by B. Birshtine, President Akaev's man. The rumors that HEMCO was broke appeared before Mr. Hunt left Bishkek. Why? Also, why didn't he obtain all the information?

"It is simple. Another clan doesn't want him because the president does. This is the very reason, believe me. And they are skillful enough using the competition among the foreigners."

(M. Parishkura, Deputy Minister of Foreign Trade; conversation with author, September 1992)

Rule 1-8

Give as much information as possible about your activities in the country to representatives of the *different* clans. Try to make them compete to become your ally.

Since you made this kind of competition rational (i.e., where profit is involved and not ancient animosities), it should be much easier to manage the situation.

◆

The First Secretary (F.S.) of the U.S. Embassy in Bishkek didn't look happy after being transferred recently from Switzerland. Visiting the newly arrived U.S. diplomat was Bolot Shamshiev (B.Sh.), the movie maker; the previously mentioned exploration and mining vice-presidents (V.P.E. and V.P.M.); and the author (A.).

◆

A.: Mr. First Secretary, please, meet Mr. Bolot Shamshiev — a Kirgiz movie director who has won European fame and is a member of Kirgiz Parliament.

F.S.: Nice to meet you, gentlemen.

And he should add with emphasis: and you, Mr. Shamshiev.

Rule 1–9

Your greeting should emphasize both the name and the position of the person with whom you shake hands.

Orientals use this greeting among themselves also but in very special cases. Make this case special.
The following quote is by J. Matlock, former U.S. Ambassador to the FSU. It was said during a reception at the U.S. Embassy in Moscow, Christmas 1988. The quote is a good example of showing honor to an esteemed guest.

"Mr. Shamshiev, you are as unique as your country, your people, and your culture. I must say that I am proud to have you among my Soviet friends, and since all of you are my wife's and my guests today, I have good reasons to think that I didn't waste the time here."

F.S.: How is your business here?
V.P.E.: I believe we have made good progress. What we want is to improve their oil production and make some exploration across a few of their basins. And they are ready to give us a share of the increased production and also of the production from newly discovered fields, if any.
F.S.: It is a long-term investment, isn't it?
V.P.M.: Yes, it is. So we want to negotiate some gold properties also. We think gold is a short-term project. It would reduce the risk we face with oil.
F.S.: And it would double the investments at the same time. . . . So you have a good impression of this country.
V.P.E.: I like it very much.
F.S.: The scenery looks like this in Switzerland, but the accommodations here could not be worse. My family has met with a lot of problems. First, the kids have no school to go to . . .

Discussing scenery and family should be the beginning of this conversation (not business). Only discuss business after asking and answering a few — 10 or more — common questions on health, weather, or family status. Mr. Shamshiev's art could have been discussed. Coffee should have been served. The first secretary missed a good opportunity to discuss his guest's problems with the right person. Mr. Shamshiev could be very interested (for understandable reasons) in presenting the problems of the foreign personnel to the Kirgiz Parliament. But — since the etiquette wasn't kept — he didn't, in spite of the available political profit he could win.

"Is he a diplomat? Anyway, he doesn't fit in Kirgizstan. We need a strong American influence here, and he defames it, creating enemies for himself. I'd like him out."

(Mr. B. Shamshiev; comment to the author while leaving the U.S. Embassy)

A month later a party took place for the presentation of an American company. Vice-President Kulov of Kirgizstan and the U.S. Ambassador were invited. After finishing a speech, the ambassador approached Mr. Kulov and said: "I am terribly sorry, Mr. Kulov, but I have to leave. My wife is waiting for me. I hope you will receive this letter tomorrow. . . ."

By Oriental customs, this was a direct affront.

> **Rule 1–10**
> **Business is nothing more than the consequences of personal relationships and etiquette.**

The business interests (and the mutual interest) are far secondary since these peoples mostly have no money interest in the deals for which they are responsible.

> **Rule 1–11**
> **Never mention your wife or daughter(s) while talking.**

Wives and daughters are esteemed in the American culture; however, in the Oriental culture they are just a tabu — not to be mentioned. If you want to insult someone, simply break this tabu. Of course, not one Kirgiz thinks that a foreigner should know the nomads' laws and customs but, naturally, a special kind of behavior is expected from a guest. The question "What is your income (or wealth)?" cannot be accepted by an American, who would consider it too personal. Does it matter that he was asked by a foreigner?

The author wasn't permitted to quote Mr. Kulov's comments concerning the Ambassador's abrupt departure.

◆

The following excerpts are from a meeting held during June 1992 in the city of Bishkek, Kirgizstan.

The meeting was with Mr. Kapar K. Kydyrov (K.K.), General Director of the State Committee on Precious Metals *Kirgizzoloto* (Kirgizgold) and his Deputy Mr. Anatoly T. Serebryansky (A.S.). The American representatives were the vice-president (Exploration) of a small American gas company (V.P.E), the vice-president (Mining) of the same company (V.P.M), and the author (A.).

◆

A.S.: I have a question: How much gold is your company producing?

V.P.M.: As a matter of fact, our company doesn't produce gold at all. But the president of the company and I have some mining experience. I was involved in a few mining projects in the United States and abroad, and, since I was the secretary of the Regional Mining Society, I have good connections with some gold companies in my country.

A.S.: So you are promoters, aren't you? You want to bring a large company here and be paid for your services.

K.K.: But, this isn't in our interest to give your company exclusive rights in our large gold fields.

V.P.M.: We aren't promoters.

It is better to say: We are promoters if you wish to call it this way, but we would invest our own money also.

V.P.M.: We have money to invest; we have the necessary experience; of course, we will have partners here.

A.S.: How large is your company? How much money do you want to invest?

V.P.E.: Well, we aren't a large company, as you can see from our brochure.

The reply to A.S. should have been: We aren't a large company on the American scale.

V.P.E.: And I cannot disclose here how much money we are willing to invest in Kirgizstan. Nobody knows how much these projects will cost.

He should know the approximate cost of each project to be discussed. He must disclose these figures here at the table. And, he should tell them what would be the portion of his company's investment.

K.K.: Do you have a ready partner willing to come with you to Kirgizstan?

V.P.M.: We haven't yet. After we make a pre-feasibility study — under the terms of this proposed protocol — we will try to sell the project to some potential partners, and we will discuss with them our portion of investments. I know exactly who the partner might be. As for today, it is enough, isn't it?

Another way of explanation needs to be chosen: Yes, we have potential partners ready to go with us, and they are awaiting the results of our negotiations with you. And, at least one of them is a giant of American industry, believe me. You understand that, first, they want to know what this deal is about. We haven't the necessary data yet, but I am sure I will bring them to develop the Kum-Tor deposits.

A.S.: Also, we have lawful doubts about giving you the data. First, it is still secret and we need the permission of our government; second, in order to find a partner, you must sell the data to him. Why can't we sell our information? Do we need a middleman?

A.: I think this issue has to be included in the protocol and then no information will be sold.

V.P.M.: Furthermore, gentlemen, I know you were the guests of Newmont just a few weeks ago, and you were very impressed, weren't you? Tell me, please, if it isn't a secret, how much time will it take Newmont to make their decisions on the Kirgiz project?

K.K.: Half a year, I believe.

V.P.M.: Then they will proceed to the pre-feasibility study; then they will complete the pre-feasibility study; then they will try to estimate their profit; then they will prepare the terms of a contract and negotiate this with you; and then they need approval of these terms throughout many levels of their company's hierarchy.

A.: You know this procedure, don't you, gentlemen? How often have you traveled back and forth between Moscow and Bishkek to have your projects approved? Fifty times a year?

K.K.: Well, we did.

A.: I want you to realize that the large, public Western companies are similar to the socialistic enterprises. They are never fast. By choosing a small, private company — even as a promoter — you are winning a lot of time, since small companies work faster.

A.S.: It would be a kind of an argument if we were in a hurry.

V.P.E.: Aren't you?

K.K.: If we are, we don't know this.

V.P.E.: Surely, you are in a hurry with the oil.

A.S.: It isn't our business.

V.P.E.: Is it your country?

Rule 1–12
Don't push your interlocutors when you face resistance.

If you do, later you may expect their resistance and obstruction, even if a direct order to meet your interests has been given by the government.

A.: Well, gentlemen, I am sad to see that we didn't reach an agreement on Kum-Tor at this time. I believe we should return to this issue later. Let us change the subject now and discuss the Jeruy and Taldy-Bulack deposits. Is it okay?

These two deposits weren't as manifestly feasible as Kum-Tor. But the problem of keeping contacts and relationships open was evident and should be solved.

Rule 1–13
Try to set up a second meeting to discuss the problems which were not solved during the initial meeting.

If you succeed, then nothing is lost. But the second meeting is the last chance; you should know exactly how to make your interlocutors change their minds.

The following conversation concerning bribery was, is, and will be possible in Russia but not in Kirgizstan:

"I will give you a million (rubles) and nobody will know it."

"Give me two (million) and let everybody know."

Notice the difference in the same type of conversation in Kirgizstan. Rumors are very important in Kirgizstan, and therefore the conversation would sound like this:

"I will give you a million and only my people will know it."

"Give me two but tell my people that it was four."

The law strictly forbids all kinds of bribery, but the ancient customs require them. The more a person is given, the more respect this person obtains. Obviously the rumors exaggerate the sum and make the person's friends and enemies envious of his success, and thus they show him even more respect. So rumors are most important.

During the last 75 years, any contact with foreigners was dangerous, even though it was required and sanctioned by the authorities. So it was impossible for foreigners to bribe local managers. However, the practice was restored in the beginning of the 1990s, though it still remains extremely dangerous in central Asia (not in Russia).

◆

The issue of bribery is a very delicate matter. It is possible to avoid bribery by the use of rumors, but you then enter into the treacherous world of Byzantine diplomacy. Be sure you are skillful and lucky.

Rule 1–14

Let rumors of your company's wealth and generosity run before you. Never use community transport (buses, trolleys, etc.) — rent cars or vans; rent helicopters to see sites (deposits and oil field areas); dress in the best style. Some exaggeration is permissible.

Rule 1–15

Let the local people pay your bills (restaurants, rents, hotels, phones, etc.) if they insist on doing it.

They pay rubles, you return their expense to them in currency (dollars). This is a big deal for them.

Rule 1–16

Ask your partners which gifts they want you to bring them. Bring the gifts.

◆

At the *Kirgizgold* office, it became clear that the two officers (K.K. and A.S.) would like to stay with Newmont. It was the first time the two had ever traveled abroad to visit Newmont's headquarters, and obviously they were impressed. Of course, rumors had spread that these two gentlemen were corrupt. This was a rumor they didn't like at all, and their real obligations to Newmont didn't matter — they must eliminate any suspicions as soon as possible. They were advised that the best way to do this would be to establish a relationship with a competing corporation.

A second factor was Siabeco's alarm at the presence of a powerful competitor (Newmont) in Kirgizstan. Since Siabeco didn't have a ready substitute to be brought into Kirgizstan immediately, they would try to insert a temporary substitute into the fight for the Kirgiz gold — until they could find a unit profitable for themselves.

The third factor was that Newmont kept the deal in a shadow of reasonable secrecy, and some powerful clans were concerned that they had been left out of the game.

Use Rules 1,2,4,8,10,12,13, and 14 to take advantage of a situation like this.

◆

◆

A meeting was held in August 1992 in the city of Bishkek, Kirgizstan with Mr. Chengyshev (C.), Prime-Minister of Kirgizstan.[4] Participants included Mr. Tekenov (T.), the Chief of the State Geological Committee; Mr. Kydyrov (Ky.), Chief of the State Committee for Precious Metals; Mr. Kholbaev (Kh.), General Director of the State Oil Company; Mr. Omuraliev, Minister of Industry; the chief geologist of an American mining company (C.G.); the vice-president of an American oil company (V.P.), and the author (A.).

◆

C.: *Zdravstvuyte.* (How do you do?)

V.P.: *Saalyam Aaleykum*, Mr. Prime Minister.

This is a traditional Oriental greeting. After the dissolution of the FSU, the relations between Kirgiz and Russians were ruined, and Russians and other aliens used this greeting to emphasize that the person(s) they have applied to does not belong to their circle. Of course, Mr. Chengyshev expected the usual English: How do you do? "Saalam Aaleykum" would have been appropriate if a Kirgiz translator was present and Mr. Chengyshev had used this greeting first. Basically, the use of Saalyam Aaleykum by the American started the interview off on the wrong foot since it was, most likely, experienced as a put-down by the Kirgiz Prime Minister.

C. (smiling):

Of course, *Saalyam Aaleykum.* I believe you miss your families, being so far from home. Please, have a seat and make yourself comfortable. You have good connections with your families, don't you? I know our phone system works badly.

---

4. Mr. Chengyshev was relieved of his post in November 1993.

V.P. Not at all. I spoke with my wife recently. Thank you for asking.

He must say: with my family or with my son, at least.

> **Rule 1–17**
> **You should hire a local interpreter for the introductory part of the conversations with VIPs.**

The local translators are not qualified to interpret technical details. But they would automatically improve your mistakes on local etiquette.

C.: Well, your families know that you are in good shape in our country. How do you like it? Have you seen the sights here?

V.P.: Oh, this is a wonderful country. The scenery looks like Colorado. Well, the peaks are twice as high and the slopes are twice as rich with gold compared to the Rockies. We visited a few places where the gold deposits are located.

This tact seems like the beginning of negotiations. Too fast.

> **Rule 1–18**
> **Spend at least the first 10 minutes of a meeting in simple, non-business conversation. Wait for the host to begin the business negotiations.**

In order to understand whether the negotiations have or have not started, you need to know the host's background.

C.: They reported to me that you enjoyed an excursion by helicopter to the Zjelal-Abad region.

V.P.: Yes, we did. We visited your oil fields within the Fergana Valley, near the point where the Uzbeks had this gigantic blow-up recently. What a disaster! They had this oil from a depth of more than 5,000 meters. You don't have a single well in the valley as deep, do you?

C.: My specialists will tell you.

Kh.: No, we have not.

Once more this is too fast at first.
Second, you should not ask an Oriental leader questions which the ambitious man cannot answer. Before this meeting started, the V.P. was told that Mr. Chengyshev is an agricultural specialist.

A.: On our way back, we visited the forests in the hilly Zjelal-Abad region. Our helicopter landed among these gorgeous nut trees. You call them Greek nuts. Why?

C.: They are really unique. Alexander the Great himself brought them from Macedonia, and now these woods occupy an area as large as all of modern Greece.

C.G.: This must also be a source of wealth for you.

C.: Not only physical wealth but a national treasure. If we had gold deposits there, I would not allow development of those deposits, believe me. But if I am told that there is oil here under my office building, I would ask you to drill right here.

Later, a preliminary agreement for exploration of oil within the Zjelal-Abad area was signed and it didn't correspond to Mr. Chengyshev's words. You have to accept that a Word is a value in itself here but not an obligation for any activities. "A value in itself" refers to a deep, ancient attention to how one's Master had expressed his wishes. It is rather from a psychoanalytic area — to guess one's Master's wishes even before the Master himself has recognized them.

Rule 1–19
Listen. Sometimes listening is boring, but the success of the negotiations most often depends on your ability to listen.

A.: I think these American representatives will establish a special fund which would help your government take care of these unique nut forests.

C.G.: Oh, yes, we are ready to do so.

Rule 1–20
Be aware that easy chatting can turn out to be the most important part of the negotiations, if the host would let you know his specific interest.

It will almost never be expressed directly. But since you have met his interest positively, you have relocated your deal into the field of etiquette; now it is his turn to be positive. He is obliged to say "Yes" regarding your wishes. He has no obligations to follow his "Yes" yet, but from now on his negative actions are restricted.

◆

The following excerpts are from a meeting with Mr. A. Akaev (A.A.), the President of Kirgizstan, and his State Adviser, Mr. L. Levitin (L.L.) in August 1992. The American participants included the chief geologist (C.G.) of an American mining company, the vice-president (V.P.) of an American oil company, and the author (A.).

◆

A.: So, my plan was to have an agreement with your government on your part of the Fergana Valley; with the Uzbek government for their portion of this valley; and an agreement with the Tazjikstan government for their portion of the valley. Then we have a large region for the workover of existing wells and for exploration. A year or so later, we will think about the pipeline. It may cross Kazakhstan and Russia toward the pipeline which Chevron wants designed to connect the Tengiz oil field with the Black Sea ports. Or it may cross Afghanistan and Iran toward the Indian Ocean, if you don't believe that your old metropolis will appear as a good neighbor for you.

A.A.: We believe that it is possible to make them so. How do you feel; is Mr. Bush going to win the election this year?

V.P.: I hope so, Mr. President. He is very popular.

C.G.: At least he intends to win. I think the American people will not try to leave the world without an experienced leader in such an alarming period of time.

A.A.: Now it is only the beginning of that. I'd say we are rich but trapped, and we need assistance. These are the consequences of our belief that the FSU was everlasting.

A.: Make Kirgizstan more attractive for foreign investments. It will solve a few problems.

A.A.: A few only, but they are important.

L.L.: If I may, Askar Akaevich? Gentlemen, recently the president has signed a decree which requires an introductory contribution of $5,000 when a foreign investor wants to begin operations in Kirgizstan. It is partly refundable if this investor later decides not to invest. I will give you the text. You will be the first in this club of foreign investors.

Of course it was his idea and he liked it.

> **Rule 1–21**
> **Try to avoid paying money to any governmental subdivisions until you have signed a contract.**

This is a way to be respected.

A.: We are honored, Mr. Levitin. I am not sure we will be the first since we were here when this decree was issued. So we really have spent our money. But I believe that Mr. President knows that we have already contributed a sum to help the Kirgiz region after this earthquake.

A.A.: No, I didn't know that. Thank you for your kind assistance.

> "And then — why should we neglect our own interests? We, who finance the gold production on Kum-Tor, have invested $1 million gratis in developing the Issykkul region, donated some money to the Jalal Abad University, and to the Bishkek Business School. But you have to agree that everything has its limits. Yes, we have signed an honorable contract; however, Cameco isn't going to spend a penny for charitable purposes anymore, until your government makes up its mind and gives us a solid conclusion on our Feasibility Study. Of course, each project might be improved once more and once more; nothing is perfect. But there is a final draw here, and nobody can transgress it."
>
> (L. Khomenuk, President of the Kum-Tor Operations Company, Cameco's branch; interview with a reporter of the *Slovo Kirgizstana*, Bishkek, March 5, 1994)

◆

Later in November 1992, Mr. Dastan I. Sarygulov, the Governor of Talas region, was appointed the general director of a newly created state company, *Kirgizaltyn* ("Kirgizgold" in Kirgiz language). He became responsible for all the contacts with foreign companies wanting to invest in the Kirgiz gold industry. Now the foreigners had to begin from the beginning. Well, some of them did.

In November 1993, some members of parliament proposed that the contract with Cameco on Kum-Tor, signed by the prime minister, has to be declared null and void. President A. Akaev in a speech to the parliament confessed that "his belief in the Swiss banking group was a mistake."

◆

"When a choice between the democracy and presidency appears, it is very easy to predict what will be chosen by an acting president anywhere. In here, my question is: Have we enough positions in the government for all the local separatists?"

(A. Zjumagulov, Governor of the Chu region, former Prime Minister;[5] conversation with author, November 1992)

"It makes a difference working within the mid-continent of Eurasia rather than within the mid-continent of Antarctica, even though it is not as easy to see, especially when it comes to Kum-Tor. But who do I have to ask for permisson to work here? Sometimes I feel that we have to apply to the Emperor Penguins."

(Manager of a large American mining corporation; conversation with author, August 1992)

◆

In the beginning of 1994, in accord with the General Agreement, Cameco submitted its Kum-Tor Feasibility Study to the government of Kirgizstan.

In February 1994, in a special session of government, President Akaev was silently absent; the newly appointed prime minister, Mr. Zjumagulov, wasn't the person who signed the General Agreement, and he had no personal reasons to support it; the Swiss banking group Siabeco had been driven out of Kirgizstan a few months before. The government declared that Cameco's feasibility study was contrary to the General Agreement since this agreement had specified the development of the total gold reserves whereas the study (made by Kilborn Western Inc.) provided the production available only from an open pit (less than 70% of the total reserves). The government decided to give Cameco the chance to change the Feasibility Study properly.

◆

5. He was reappointed as Kirgizstan's Prime Minister at the end of 1993.

"We made our Feasibility Study for the first phase, when the ore was produced from the open pit. It is absolutely important for obtaining bank credits as soon as possible. We'll produce the ore over 12 years. Nobody can predict what will happen then. We must begin making money as soon as possible. Each day of delay means half a million dollars. Kirgizstan needs money now. Does your government and parliament understand? By the way, in no other country does the parliament meddle with such business. I think President Akaev has said the same. . . ."

(L. Khomenuk, President of the Kum-Tor Operations Company; interview with a reporter of *Slovo Kirgizstana*, Bishkek, March 5, 1993)

What a mistake to tell this to the mass media!

"The foreigners will skim off the cream, and then we will have to dig a mine in order to produce the third portion of Kum-Tor's gold. We are outdated people, but we aren't stupid. He says that Cameco has spent a lot of money. It is up to them. They will demand a forfeit; they may try. What is the reason for such demands? Nobody approved their feasibility study."

(K. Idinov, Ph.D, Chairman of the Standing Commission for International Relations of the Kirgizian Parliament; phone conversation with author, April 1994)

Chapter 2

Uzbekistan: Wages Talk

◆

They were merchants on the Great Silk Road long before the time when the Romans tried to conquer the Albion; long before Alexander the Great came to introduce himself here; and even before their Shah (Darius) gave permission to the Jews to return to Jerusalem and restore their temple. No other people anywhere in the world have such an ancient tradition to be merchants. So be prepared that they will understand your interests before you do.

The population of Uzbekistan is almost 20 million people, and 66% of them had declared themselves as Uzbeks in the time of the FSU, when it became profitable to be an Uzbek in this country. Ninety-five years ago, 81% of the population had declared themselves as Tajiks, since it gave them some advantage under the laws of the Russian Empire. The Uzbek language is a branch of the Turkic family, and the Tajik language is a branch of the Pharsee (Persian) family. The two languages are totally different, and the majority of the population speaks both languages.

The Uzbeks believe that they are Sunni Muslims; the Tajiks are Ishmaelites (a very specific secret Muslim sect). The two branches of Islam fought for over a thousand years. When the Sunni rulers won here, some Tajiks became Sunnis (in public only); when the Shiite rulers won, the Uzbeks converted (in public only).

In 1980 about 6% of the adult population of Russia were members of the Communist Party; in Ukraine, more than 8%; in Moldavia, more than 9%; in Uzbekistan, more than 13%; in Armenia, more than 14.5%. These figures might be a reflection of the ability of different nations to adapt and their will to survive.

The name Uz-Bek means: I am a Bek (Baron) for myself. After some Turkic tribes conquered the region along the southern edge of the great central Asia desert of Kyzyl-Kum (Red Sands) and along the two great rivers Amu-Darya and Syr-Darya, they settled in the countryside while the Tajiks continued to live in the great and ancient cities such as Hiva, Samarkand, Bokhara, etc. Stalin decided that this was wrong, and he changed the status quo. The ancient cities populated by Tajiks were included in Uzbekistan, and the arable lands populated by Uzbeks were included in Tajikistan. Now Uzbeks became rulers of the Tajic population, and Tajics became rulers of the Uzbek population. So the only support the local rulers could possibly obtain came — from Stalin from Moscow — from the KGB.

The two nations are as ancient as the rules of these political games. During the final years of President Gorbachev's rule, the bloody incidents[1] in the Fergana Valley, Osh, and Samarkand regions were stopped by the people themselves.

◆

> "We don't believe them and they don't believe us. If they will play Fundamentalism, we shall play Pan-Turkism; if they will not, we shall not. But they will and we shall, if they have much more money than we or we have much more money than them. Only the state can remain a regulator in this republic. We must keep the system we inherited. 'Support to the State and to the President' — this is the only and the last point of view the two nations could share together. No status quo is wrong if it means *peace*. The time for a war hasn't come yet."
>
> (Professor B. Tursunov, Chancellor of the First Uzbek Independent University of Business and Diplomacy; conversation with author, New York, October 1993)

> "You were a guest in my home, you saw my father. The grandfather of his father was born in this house. . . . I believe the Uzbeks are smart enough to be friendly with the Russians here. I speak their language as easy as Russian, and my grandfather could write using all three alphabets they used during last 70 years (Arabic, Latin and Cyrillic). Now the Russians become a great stabilizing factor here in spite of the terrible nationalistic pressure we feel. Do they want us to leave? Those who have nothing to lose, they do. . . . Who will manage the industry? It isn't for Americans because it is rather old. . . . You know,

1. They say that these incidents were set by the KGB to prove to the local population that it cannot survive without Moscow's rule.

there is a limit for status quo, and if they are smart enough to keep it, we are ready to stay here."

(V. Zolotukhin, President of Intersystems, former officer of the Soviet Army, former member [from Uzbekistan] of the first and the last Soviet Parliament; conversation with author, Tashkent, August 1992)

"Privatization, democratization—words. Suppose a few members of our parliament came to the United States and held a speech like this: 'You guys should enjoy life since you are rich enough. Two wives are better than one, and four wives are twice as good. We have had this experience for a thousand years, believe us and change your minds immediately.' Polygamy. Words. You would buy them their tickets back the next day, wouldn't you? The U.S. senators behaved here like a few bulls in a china shop. See, the East is a very sensitive matter; now Uzbekistan is especially brittle, frail, and fragile. So you better stay away with your brilliant ideas. Come and make your money here and don't forget that we badly need money too."

(V. Tavorovsky, General Director of *Uzbekneftegazstroy* [Uzbek Oil and Gas Pipelines Installation]; conversation with author, September 1992)

◆

In March 1992, a failure and blowout at well #2 (Minbulack) brought a disaster to the Fergana Valley and proved that within this valley the deepest intervals of Neogenic deposits are oil-saturated. An enormous rate of flow (10,000 metric tons per day) could be observed for almost two months. The bottomhole depth was as much as 5,326 meters and the initial pressure at the well mouth reached 1,067 atmospheres; therefore, all three blowout preventers (BOPs) flew out at a distance of one-half kilometer from the well. Neither the casing nor the BOPs were designed to hold such pressure. Since the oil was light (specific gravity= 0.798) and mobile and since the Syr-Darya River's tributary was at a distance of less than 0.7 km, the blowout could have become a national tragedy if the oil stream reached the river. This river is carrying 45% of the potable water of Uzbekistan and 55% of the water for irrigation of the fields in the valley and in the Kyzyl-Kum desert.

An American businessman was visiting Bokhara and Samarkand during this spring. Coming back to Tashkent, he managed to see the Minister of Foreign Trade of Uzbekistan, and the two gentlemen signed a contract. Under its terms, the American was obligated, at his own expense, to bring a team and the equipment from the United States to fight the fire and to shut in the well. In exchange, the American received exclusive rights for exploration and production within the Uzbek portion of the Fergana Valley.

Neither the artillery shells and air bombardiers nor the mullahs could stop the fire at the well until the walls caved in during late April. The local people say that a very respectable mullah had scattered some salt along the oil-lake dam, and this dam became strong enough so that no oil poured out from the lake into the river.

The American team and equipment came late, and the company was never paid by the American who initiated the contract. He tried to sell the Fergana project to different American and Canadian oil companies but none of them accepted his conditions. The Uzbek government tried to declare their contract *null and void* but wasn't ready to bring the deal to court. These circumstances postponed the oil rush in Uzbekistan.

◆

◆

During 1992, Uzbekistan produced 42.5 billion cubic meters of gas and planned to produce 51 billion cubic meters during 1993. The gas export was 5.5 billion cubic meters in 1992 and was expected to be 13.4 billion cubic meters in 1993.

More than 2.5 million metric tons of oil were produced in 1992 and less than 2.3 million metric tons in 1993. This republic consumes up to 11.9 million metric tons per year, so the import of crude oil is about 9 million metric tons each year. They ship their cotton and gas to Russia in exchange for oil. The two refineries (in Fergana and Altaryck) are very old, uneconomical, and insufficient even though their capacity is up to 8.5 million metric tons per year of crude oil.

The potential oil resources at the Minbulack field (Fergana Valley) are up to 875 million metric tons (in-place); the depth, up to 5.5 km. The potential oil resources at the Kubamalakma field (Surkhan-Darya depression) are up to 790 million metric tons (in-place), depth, up to 3.3 km—sulfur content, 3.1%; condensate, up to 31%. These are the short characteristics of the two largest oil fields of Uzbekistan.

Besides these, there are seven other oil fields with possible resources of up to 1.1 billion metric tons in-place in the Fergana Valley; and, an additional four oil fields with possible resources of up to 0.7 bln metric tons in-place total in the Surkhan-Darya depression.

The central and western parts of the Fergana basin are poorly explored because: (1) it is deep (down to 9.0 km) and the tectonical structure is very complicated, and (2) the irrigation channels (*arycks*) on the surface make it inconvenient for the seismic survey.

Three other basins in Uzbekistan are less promising. The Bokhara-Hivinskiy basin produces mainly gas; the Ust-Yurt basin is the most shallow basin (down to 2.5 km only), but it is much less explored. The Karshy basin produces mainly gas, but the estimate resources of oil are as much as 0.4 billion metric tons in-place.

There are about 1,500 deep wells in the Fergana basin, but less than 800 of them produce oil. The reason for this is the poor quality of mud, cement, casing, and design. Some examples of this follow.

It takes as much as 1.5–2.0 years to drill a deep well here, and a deep well costs as much as $2.5 million.

The load capacity of a drilling rig is 400 tons — 120 revolutions per minute of rotor; two mud pumps: 84 cu ft/min, 1715 psi and 38 cu ft/min, 3571 psi.

They use water-based mud, three-stage mud cleaner, imported casing, tubing, and drillpipes; hydraulic tongs; BOPs and christmas tree rated to 15,000 psi.

The wellhead pressure is expected to be about 12,800 psi. The 20 lb/gal mud weight was used for penetration of the Minbulack oil reservoir. The 19.5 lb/gal mud weight was used for penetration of the Gumkhana oil reservoir. For maintenance of mud, they use treating chemicals from Russia, Crimea, and Uzbekistan. They use barite for the weighting of mud and cement.

The working characteristics of the cementing job are:

Ca	320	4600 psi	260 h.p.
A	400	5800 psi	450 h.p.
AN	700	10000 psi	450 h.p.
CF	1000	15000 psi	350 h.p.

Of the 10 deep wells drilled on the Gumkhana field, none was completed since the tubing was obstructed with sand. The testing was released by open flow. Obviously, the tubing and casing were collapsed. One well was equipped with a slotted liner; the width of slots are 1.5 mm. Daily output was 70 tons of oil. The attempt to perforate wasn't successful.

Thus, about 700 deep wells in the Fergana Valley are waiting to be worked over, and some additional 1.1 million tons of oil per year (more than 8 million Bbl/year) may be produced. But for now, Uzbekistan imports her crude oil from Russia, as was mentioned above, using railway tanks. There is no oil pipeline connecting Uzbekistan with a sea port.

◆

◆

Currently, the real cotton and real gas — no papers and reports — should be sent to Russia in exchange for oil, and this is painful for the economy of this republic. Before *perestroyka,* a report of a harvest of over 6 million tons of cotton was sent to Moscow each year. Though the price of cotton was low enough, the price of oil was comparatively much lower.

After this report had been sent to Moscow and all the money and decorations received from Moscow, a few fires would be set at the cotton bins. Where 100 tons of cotton were destroyed by the fire, the officials would report that 10,000 tons of cotton perished. Hundreds of such fires destroyed thousands of tons of cotton, but in the reports, the figures of hundreds of thousands were shown. Then, if a thousand tons of cotton were delivered to the factory, it was accompanied with documents reporting a 10,000 ton delivery. And the receiver signed these documents because of a bribe of thousands of rubles. Often a fire was set at a factory storehouse to cover the deal, and, again, the figures in the reports had no relation to the real amount of the burned product.

An Uzbek peasant was paid less than 200 rubles for collecting 2 tons of cotton per month. Using the ploys above, the Uzbek government was paid about 25,000 rubles per ton of cotton (officially the price per ton was 609 rubles). However, the government never used any of this money to improve the environment and healthcare of the local peasants and the inhabitants of the areas surrounding the Aral Sea.

◆

Water is diverted from the region's rivers. Since 1960 this has shrunk the surface area of the Aral Sea more than 40%. The use of pesticides and other chemicals pollute drinking water. They say mother's milk may be contaminated. The local people have suffered from throat cancer and high infant mortality.

When Yuri Andropov, the former chief of KGB, became the general secretary of the Communist Party of the Soviet Union (CPSU) and began to dig into these deals, he was admitted to the hospital for the first time. They quote him as saying: "Here are the limits of the KGB power, and thus — I mean what I say — the end of a great state."

In the time of M. Gorbachev, the former deputy minister of interior affairs of the FSU — Leonid Brezhnev's son-in-law — spent five years in jail for taking bribes in a few episodes of the *Uzbek deal.* Mr. S. Rashidov, the First Secretary of the Central Committee of the Communist Party of Uzbekistan died under unclear circumstances. Three first secretaries changed one after another during less than four years, and each of them spent a term in jail even though M. Gorbachev tried to forget the deal. Then Uzbekistan became independent; a monument to Mr. Rashidov replaced Lenin's monument at the central square of the city of Tashkent, the capital of Uzbekistan, and the square itself was renamed also. Most of the suspects of the *Uzbek deal*, who didn't perish in prisons, were released but never returned to their former positions.

Until the year 1989, the Uzbek government paid only 8.3 rubles ($3) for a ton of crude oil delivered from the West Siberian region; in the year 1992, 31,000 rubles ($31). And it became harder to earn this money. The fuel supply shortage affected all industry and agriculture.

In the summer of 1992, while flying between Bishkek and Tashkent, a few Americans suddenly found the airliner in the center of a huge thunderstorm among the peaks of Tien Shan. A noisy fire-show outside the aircraft was accompanied by a dangerous vibration inside the plane, and a few passengers were seriously wounded. Since no stewardess was onboard, they remained helpless until landing. By the rules, pilots should outflank the zone of thunderstorms, but not having enough fuel for such a maneuver, they had no choice but to cross directly through the thunderstorms.

A few weeks later, the Americans were on another local flight from Minbulack to Tashkent. Just as the wheels touched the runway on landing, the engines of the AN-24 (a small Soviet liner) stopped because it had used up all its fuel. The travelers were very shaken.

In the end of 1992, while hunting in the desert several hundred miles from Tashkent, we approached a tiny aircraft which belonged to the Republican Air Ambulance Service.

"One of my tanks was filled with water," the pilot explained.

Rule 1–22

Avoid using the regular local flights. Rent a car with a driver. Rent a helicopter or another aircraft when you are in a hurry. It is cheap enough and, of course, much safer. If you pay currency, they will find fuel.

◆

In Uzbekistan nobody knew how much gold was produced during the era of the FSU. The approximate estimation is up to 1,500 metric tons. This gold was produced mainly from the Muruntau deposit which was one of the world's largest—located in the middle of the Kyzyl-Kum Desert (Fig. 2–1). In 1992 Newmont had an agreement to rework the tailings which contain up to 1.5 grams per ton of gold.

◆

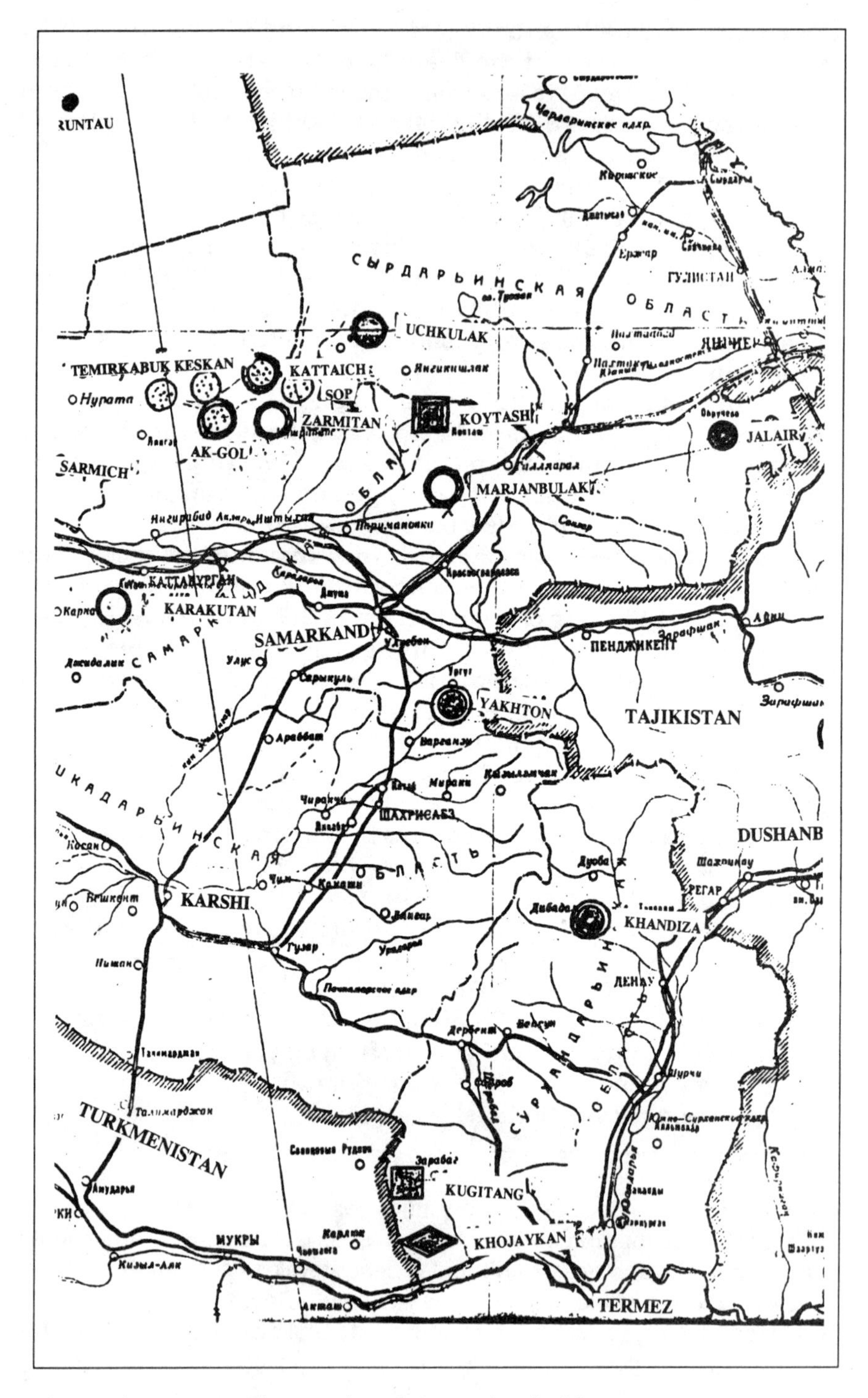

Figure 2–1 Gold Deposits of Central Uzbekistan

> "The terms of this project and the swift manner in which it came together indicated the good faith of the Uzbek government and its commitment to encourage profit-seeking foreign investment."
>
> (R. C. Cambre, CEO, Newmont; interview with a U.S. newspaper reporter, December 18, 1993)

It took almost two years for Newmont, the largest producer of gold in North America, to obtain an agreement with the European Development Bank, which signed a $105 million loan for Denver-based Newmont Mining Corp.

> "This transaction represents the first mining project for the bank, the first syndicated financing for the former Soviet Union, and the first limited recourse private sector loan for Uzbekistan in any of the capital markets."
>
> (J. de Larosiere, President, EBRD; interview with a U.S. newspaper, Dec. 18, 1993)

There are a few other smaller gold deposits which altogether contain less gold than Muruntau; nevertheless they are objects of interest to different American, Canadian, British, Turkish, and Iranian companies.

The Turkish company BMB was introduced in Kirgizstan and Uzbekistan by the former president of Turkey who influenced the Uzbek government to approve the deal. BMB presented two addresses: one in Turkey, and another in New Jersy. When checked, the New Jersey address of their headquarters appeared to be . . . a private house.

Two gentlemen, one British and another Russian, who said they represented a British mining company, proposed the financing for development of the Zarmitan gold deposit. Some papers were signed, and then a communication came from Moscow that said the two gentlemen had tried to buy old Soviet equipment for heap leaching in Jakutiya (a republic in Siberia) for this project. This kind of equipment never worked because the technology proved to be wrong.

Representatives of a Canadian mining company collected detailed geological and mining data on the Kyzylalmasay gold deposit under the terms of an agreement. Their agreement includes supplying the Uzbek specialists with information on modern heap leaching technology and on the Western firms who manufacture this equipment. None of these representatives ever appeared again in Uzbekistan and neither did this information. A copy of the letter from this company with proposals to different Western mining enterprises to buy the Kyzylalmasay data was lying on the table of the Deputy Prime Minister of Uzbekistan, Mr. Khackulov, when I visited him.

> "We are overwhelmed. They want to work here; they are welcome; but, first, they have to be audited, and they must pay for that themselves. I have signed an agreement with National Westminster Bank of Scotland; they will take care . . ."
>
> (K. Khackulov, Deputy Prime Minister of Uzbekistan; conversation with author, September, 1992)

> "Our government is naive and has no money. Under their agreement with NatWest the bank is to be paid for its *references* by the company being verified. The references from County NatWest depend on the sum paid to them by the verified company, will they not? It isn't too clever. Our government simply doesn't know what to do with these companies which keep bringing dubious proposals. But tell me why this well-respected international firm agreed to undertake this job under such conditions?"
>
> (M. Kononova, Editor, *Uzbekiston—Contact* [monthly magazine]; conversation with author, September, 1992)

Some rich mineral deposits in Uzbekistan are described in the following information.

The Khandiza polymetallic deposit (Fig.2-1) is located 60 km from the Uzun train station and 55 km from the Denau station. The Dushanbe-Termez road is also 40 km distant. The relief is very broken with absolute marks ranging from 1,100 up to 3,000 m, but, since the ores are located close to the surface, the open pit technology may be used easily (maximum depth is less than 150 m). The field is, by FSU standards, fully explored. Total reserves are 24 million tons of ore. The following table shows the metals reserves and potential production rates.

Metals	**Reserves**	**Production**	
	(tons)	(rate)	(T/Yr)
Lead (Pb)	789,600.0	3.29%	39,480.0
Zinc (Z)	1,586,400.0	6.61%	79,320.0
Copper (Cu)	204,000.0	0.85%	10,200.0
Cadmium (Cd)	9,840.0	0.041%	.492
Silver (Ag)	2,736.0	114.0 g/T	1.36
Gold (Au)	8.4	0.35 g/T	.42

The lead, zinc, and copper reserves contain no impurities.

The Zarmitan gold deposit (Fig.2–1) is connected to Samarkand by an asphalt road at a distance of 120 km. The deposit is situated among the southeast exo- and endocontact section of a large intrusive massif. The ore body is volcanic and clastic rocks of the Silurian period and granitoids. The gold originated with lava flows through the tectonic faults. There are more than 50 producing bodies, which according to geological type, can be classified as low sulfide, quartz veins and zones of small ore intrusions. The morphology of these bodies is in a plate form, and the bodies are distributed in subparallel or overlapping order. Often they brunch out and join together and are slopes 70–80 degrees in a north and northeast direction. The bodies are up to 1,200 m, and their thickness is 0.5–0.8 m. They have been noted as deep as 800–1,000 m. The assumed depth is more than 1.5 km.

In the region, there is a mine producing ore at 300,000 tons/year, and the enriching is done at the Margjanbulack Enriching Plant about 90 km away (Fig.2–1). The established, producible reserves today are 20 million tons of ore with gold reserves of 210 tons.

The estimated reserves of gold confirmed by the drilled wells (cores) are 44 million tons of ore and 460 tons of gold, which is the reason the capacity of the mine can be increased from 1.0 to 1.3 million tons of ore and a new enriching plant built.

The method of producing ore from the deposit is underground mining. The processing of the ore is by the cyanide gravitation method with the gold extraction of 90 to 93%.

The estimated reserves of gold confirmed by drilled wells (cores) at seven other deposits of the same intrusive massif are 41 million tons of ore and 505 tons of gold in total. These deposits are: Uchkulach, Kattaich, Sop, Keskan, Temirkabuk, Akgol, and Sarmich (Fig. 2–1). Among four other discovered gold deposits in central Uzbekistan only two (Mardjanbulack and Jalair) produce ore. Deposits Karakutan and Yakhton (Fig. 2–1) were discovered recently.

A few gold deposits were discovered in eastern Uzbekistan also (Fig.2–2). The gold deposits Samarchuk, Mezjdurechye, and Kayragash reserves are 89 tons of gold in total; the reserves of the exploited deposit of Koch Bulak are 71 tons of gold.

The Kyzyl Almasay gold deposit is located approximately 120 km southeast of Tashkent (Fig.2-2). The three ore bodies occur in steeply dipping shear zones cutting Lower Paleozoic sediments and volcaniclastics. The deposit is one of several in a highly mineralized zone 10 km long and up to 350 m wide along the Alma Say Fault Zone.

Mineable reserves, assuming a 5% extraction loss and 10% dilution at zero grade, are as follows:

#1 Orebody: 10,93 million metric tons @ 5.80 g/T; 63.4 tons of gold.
#10 Orebody: 6.32 million metric tons @ 10.0 g/T; 63.2 tons of gold.
#11 Orebody: exploration incomplete, no mineable reserves.
Total: 17.25 million metric tons @ 7.34 g/T; 126.6 metric tons of gold (0.214 oz/ST; approximately 4.077 million ounces).

There is a pilot mine currently producing 100,000 metric tons of ore per year at an average grade of 5.2 g/T (0.152 oz/ST). The mining method they use in Kyzyl Almasay is very typical of all the Uzbekistan's ore deposits (Fig. 2–3).

Ore from this pilot mine is hauled by truck approximately 6 km downhill to the Angren Mill (Fig.2–2). The mill products are shipped to a processing and refining plant at Almalyk, located 35 km west of the Angren Mill (Fig. 2–2).

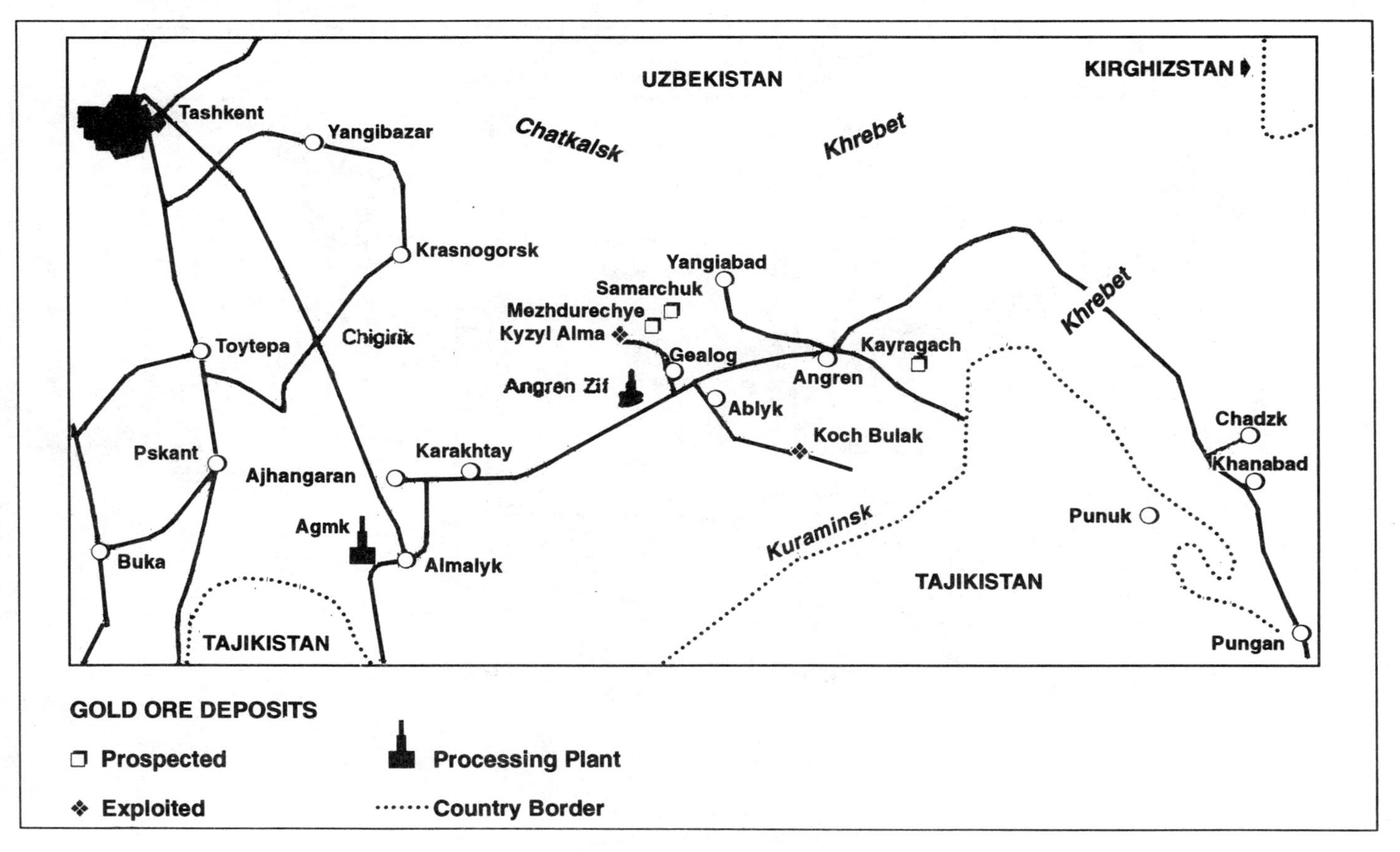

Figure 2–2 Gold Deposits of Eastern Uzbekistan

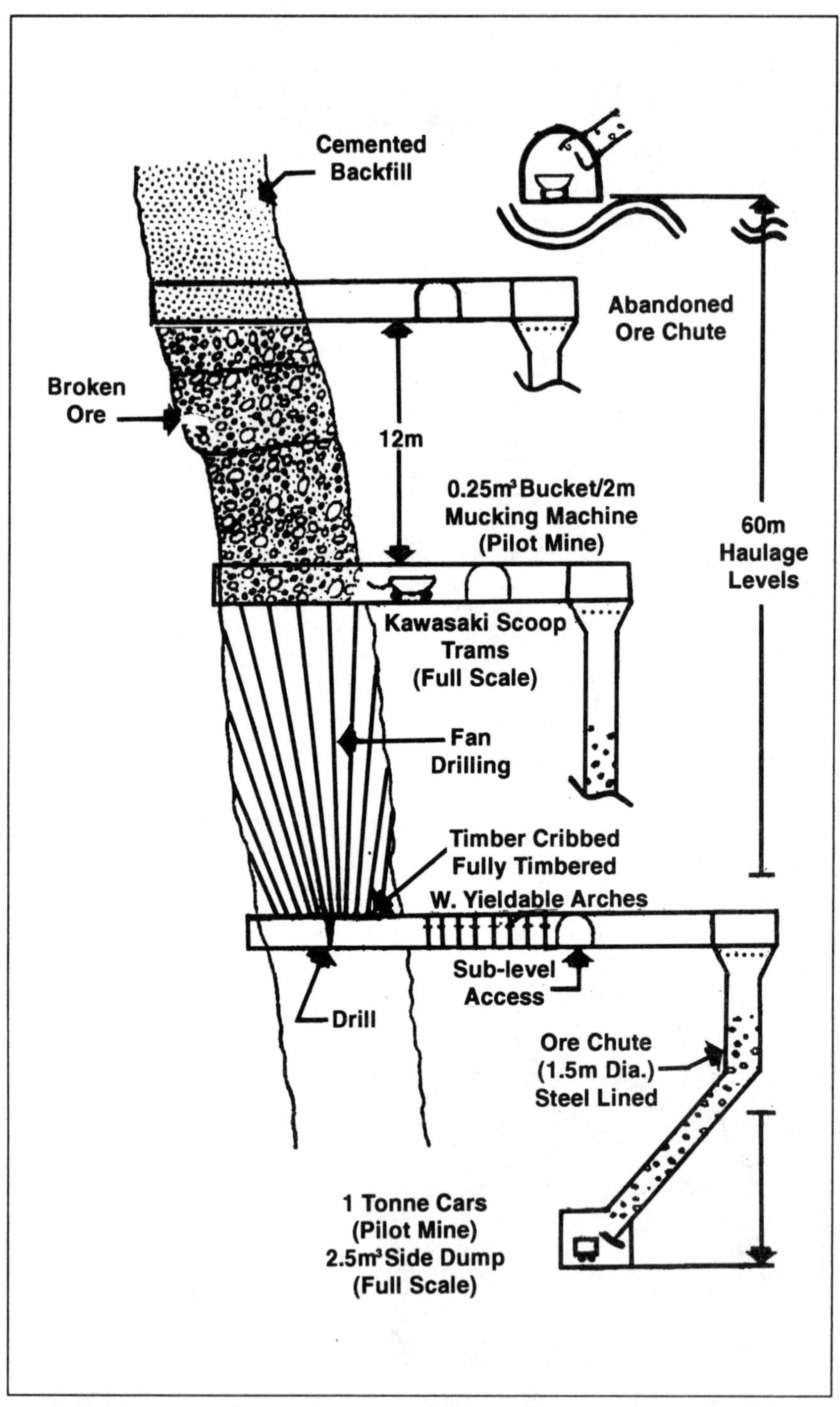

Figure 2–3 Kyzyl Almasay Mining Method

The following excerpts are from a meeting with Mr. S. Najimov (S.N.), Chairman of State Committee for Precious Metals of Uzbekistan. Participants included the vice-president (V.P.) and chief geologist (C.G.) of an American mining company, and the author (A.). The meeting occurred in August 1992 in the city of Tashkent.

S.N.: So you understand, gentlemen, that I am collecting only the foreigners' proposals. The government will decide which Western company is the best partner for our gold deposits. Then they will submit the case to our president.

C.G.: Well, Mr. Najimov, don't you think we have to make our decisions, also? We have our plans, and we want to know whether Uzbekistan is included. We obviously need data for the pre-Feasibility Study. What about that?

S.N.: I believe you have made your choice since you are here, haven't you? I can repeat that all the materials on the gold deposits are still secret. The data I gave you aren't.

C.G.: You understand, they are absolutely insufficient.

S.N.: I cannot help you. You want to help Uzbekistan, take a risk.

V.P.: I am sure you are mistaken, Mr. Najimov. We are here to make money, not to help Uzbekistan. Working here we will help only a limited number of our employees, at least.

He was right.
Entering Tashkent from the south by a highway, next to a traffic police post, you can see a monument next to the remains of a car destroyed in a huge accident. The monument reads : "He was right!" (The driver who died in the accident had violated none of the traffic rules.)

Rule 1–23
Use, as often as possible, the expression: "We are here to help you." Believe it.

You'll be right, without *any accidents.*

In the beginning of 1994, the rules of the gold game were changed.

By the decree of President Karimov, the gold industry was *privatized.* From now on, 50% of shares of the gold enterprises belong to their personnel; 33% to the state and 17% are to be sold in the free market. A new version of the *gold rush* began since the enterprises became free to choose foreign investors and partners. Under the terms of this decree, foreign investors will be paid in currency for produced gold. The current price of gold is to be equal to that on London's Gold Market.

◆

In April 1993, in the city of Tashkent, a meeting was held with Dr. A. Rakhimov (A.R.), First Deputy and Chairman of the State Concern *Uzbekneftegaz* (Uzbek Oil and Gas). The participants included the vice-president (V.P.) and the financial director (F.D.) of an American petroleum company, and the author (A.).

◆

A.: Doctor Rakhimov, last summer when we met in Minbulack, you wanted to repair and work over this world famous well which blew up a year ago. Were you successful?

A.R.: Not at all. I believed it could be done. But you know how it is here: no supplies in time, no pieces in time, no cement in time. . . . The government wants oil and tells me: You are the leader, you have to foresee, to provide for every eventuality. . . . But who can do it? Nobody.

V.P.: We can. We are here to help you.

A.R.: Oh, yes. Uzbekistan needs your help badly. If you are ready to help, you may have concessions here. I'd like to discuss it.

V.P.: Doctor Rakhimov, you have 17 deep wells at Gumkhana, a few at Minbulack, a dozen at the other fields. Each of them costs millions. None of them produce oil. Isn't it time to radically change the design, the equipment? We are willing to bring ours.

A.R.: You are welcome, as I said.

F.D.: We are ready to undertake the geological risk and to come to a production sharing contract with you. Our profit becomes dependent on the results of working over the old wells and drilling new ones. We find it impossible to transport our portion of the oil from Uzbekistan to a seaport and then to sell it at the world market rate without suffering a huge loss. Is the government ready to buy this portion and pay for it in hard currency?

A.R.: I am not sure. It is a subject for further negotiation.

F.D.: Okay, the next question is whether or not the government is willing to pay for the equipment we will bring in here. It is a guarantee against any political and other risks which may arise, as you understand.

A.R.: I asked them and I was told that they have no money.

V.P.: Doctor Rakhimov, did you explain to them your situation? If we don't come, you are doomed to failure after failure. You are condemned to spend enormous amounts of money for nothing. You have no choice. Did you explain this to the bureaucrats in your government?

How could he dare? If he tried he would be fired immediately. The next one in this position would do the same, the amount of non-productive wells would be doubled, etc. The position of Dr. Rakhimov was endangered after his unsuccessful attempt to get oil from the Minbulack deep well and now he wanted to bring a Western partner into it in order to improve his personal position in the government. If you can read this situation, your advantages may become great.

Rule 1–24

Search the firmness of the position and the situation around your main partner. If he isn't a newly appointed person (as Mr. Najimov was), he may well need your friendship more than you need his.

A.R.: They aren't interested in the explanations. They are interested in oil. We all are. If you want to help Uzbekistan, come and work; if not, it is up to you.

"I have nothing to explain to my president. He explains to me. He gives orders and our specialists obey. Nothing more."

(K. Khakulov, Deputy Prime Minister, Chairman of the State Concern *Uzbekneftegaz;* conversation with author, June 1992)

◆

There is an ancient custom in Uzbekistan's cities and villages: each week another family and its relatives prepare a huge pilaf for the neighbors on their block. This party for neighbors *(Hoshar*[2]) begins at 5:00 A.M. each Friday even though, under the communist rulers, Friday had become a business day. Those

2. Hoshar means the community's men meeting to chat, discuss problems, build a house for a young family, rebuild a dam after a disaster, etc.

neighborhoods where the elite lived have a few special families who prepare some 100 kg or more of the best pilaf for the *Hoshar*. You can be sure that even the prime minister with his bodyguards would be present that very morning each week since this is the real government of the country. One word which has been said here in the open air is worth hundreds of letters and contracts and agreements signed in bureaucratic offices.

The #1 person (the first secretary, the president) would never appear in this place; on the great holidays he would rather appear in the poor neighborhoods. It is certain that a dozen reports about the events on that *Hoshar* will be lying on his table later the same day.

For eight months the Japanese were negotiating a large refinery project in Uzbekistan, proposing both the financing ($1.5 billion) and the construction. Dozens of documents were signed with different ministers, but nobody was sure that the project would survive. However, the very next week, after some Japanese representatives appeared at the *Hoshar*, they were invited to the president's palace and the deal was approved in general. I was told that the president had to wait half year for the *Hoshar's* opinion, and that this opinion cost a lot of money.

Such *Hoshars* of different levels are enacted widely in Tashkent. I was lucky to visit one where I could ask for the data on some gold deposits. We have obtained these data.

"Of course, these data were secret under the FSU system. Now we are independent and the old regulations don't work," as it was explained to me.

Rule 1–25
Your local agent(s) should have access to the *Hoshars* at different levels (including the elite's neighborhoods).

And it is better if this agent has enough power to invite you there. He will charge you $5,000 or even more. It is worth paying this bill.

Rule 1–26
No women with you at *Hoshars*. No exceptions.

The following *Hoshar* conversation is with a deputy prime minister (DPM) and the author (A.). It occurred in March 1993.

A.: As you know, I have brought invitations from the governor of Colorado and the governor of Nevada for you and your prime minister to visit the United States.

D.P.M.: Oh, yes, I was told, thank you very much. I hope we can work together and I will come.

A.: Both Mr. Najimov and Mr. Rakhimov have advised us to submit these invitations through the U.S. Embassy to your Foreign Affairs Ministry. Is it the best way?

D.P.M.: I cannot see a better way.

A.: Then it will go its way. May I have some more pilaf, please? It is absolutely delicious today, isn't it?

The rules of Oriental talks require a pause after discussing each single deal. If you follow these rules from the beginning, this will oblige your partner to explain his negative answers. Sometimes it is very important to know his reasons, which would never be expressed if you used the Western manner of talking.

D.P.M.: Oh, yes it is. Today Selim is in charge, and he is the best, believe me. I gave him this apartment in the building next door. He had 11 children and I helped the family obtain this three-bedroom apartment. Now they have 16—even though two of his children died last year—and, of course, the family needs more space badly. It is difficult to buy a house here in Tashkent, and houses are extremely expensive. He wants to leave for the countryside, but who will prepare the pilaf?

A.: He could come once a month or so.

D.P.M.: Well, he is very special to all of us. When his children become older and the time for their marriages come, we will help them also. A few years more—these few years will be very important for our republic. And for those Westerners who will stay with us during this time—they will make money.

A.: You can see better, but in my point of view, this is risky politics. You lose time.

D.P.M.: This is all we have. What we don't have is money. . . . And these Americans you have introduced here, they are small and hungry. You suggested that I check them out with my Kazakh colleague, didn't you? I like good advice and I did.

He is a friend of these Americans; he knows them. And they are small and hungry and greedy.

A.: You know how it works there: if you have substantial money and a good name and some reliable projects, you may have a huge credit line from the banks. They are able to make things happen rather than a large, well fed and lazy entity.

D.P.M.: Well, let them come for oil. I will support it.

A.: The gold is the guarantee.

D.P.M.: If they have gold they will do nothing for oil. Nobody would choose the long-term investments (oil) if the short-term investment (gold) possibility is in his pocket.

A.: Then you pay for the drilling equipment. This is another kind of guarantee.

D.P.M.: Concession is a guarantee, isn't it?

A.: Yes, it is—if it is located along a sea beach, where a carrier could approach. You are a mid-continent country, aren't you?

D.P.M.: We are the hub of the universe. And do you know what: for us, those who cannot catch our navel aren't worth a brass farthing.

A.: Russia could.

D.P.M.: Can Americans get Russia's navel?

A.: They don't think that they need to. Nazarbaev and Niyazov[3] are more brave, aren't they?

D.P.M.: Do you want some more cognac?

Never could such a conversation happen in his office!
He didn't mention the cost of the equipment. It means that he isn't interested in playing a very old game where the seller names a price which is much more than the real price. After receiving the inflated price, the seller of the equipment would share the head *(money above what seller wanted) with local officers who made the deal happen. This game is often played when government moneys are involved.*

◆

Under President Niyazov, Turkmenistan has made efforts to become attractive to foreign investors.

The country annually produces almost 5.5 million tons of oil from the seven oil fields located on the eastern coast of the Caspian Sea. The total capacity of the two old refineries in Chardzjou and Krasnovodsk is less than 3 million tons

3. Mr. Niyazov is president of Turkmenistan.

per year, but their supply of fuel is sufficient for the Republic's demands. Thus, Turkmenistan offers almost 2.5 million tons of crude oil per year for exports. This oil is delivered to Byelorus and Ukraine in exchange for their foodstuffs. In the last few years, Turkmenistan faced the fact that neither Byelorus nor Ukraine could pay for this oil (and gas). In 1993, a project to construct a new railroad between Turkmenistan and Iran had been negotiated and both sides have started this work. When it is finished, an access to the Persian Gulf will be open for Turkmenistan's oil.

The source of Turkmenistan's main richness is natural gas. The recoverable reserves of gas in the southeast part of the Kara-Kum desert are close to 48,000 billion cubic meters. The gas pipeline network which starts here crosses Uzbekistan, Kazakhstan, and Russia before reaching Europe.

◆

> "We would prosper if Americans managed to open Iran for themselves once again."
>
> (M. Ataberdiev, Deputy Minister of Fuel of Turkmenistan; conversation with author, Tashkent, May 1993).

Rule 1–27
Never fire your local agents.

Think before promising a local agent commission money. And, you should strictly carry out your promises. Keep your local agents neutral, out of the deals, if you don't need them any more. Whatever you do, don't fire them because they will surely become your enemies.

An American company hired local agents, and they turned out to be unreliable—no cars on time; no helicopter (which was promised); no introduction to the right people (which was promised also). The American company fired these local agents by sending them a written notification.

During the next few nights, the phones in their rooms rang every half-hour. Prostitutes offered their services all night long and the following night also.

The beleagured Americans appealed to the hotel's manager. "Use them," was his answer. They appealed to the police officer.

"Invite them into your rooms and we'll try to catch them," was the policeman's answer. Of course the Americans didn't, since they were afraid of all kinds of ramifications. They had to leave this comparatively good hotel for a worse one.

However, the worst was yet to come. The general director of the plant they were negotiating with suddenly became ill and disappeared for a week or more. While waiting for the plant director, each American paid $120 per day for the hotel. After they counted their expenses, they understood that it would have been cheaper to pay the promised sum to their former local agents.

Rule 1–28
Never follow the girl you have met to her place. You may be falling into a trap.

Tashkent had been famous as a criminal city during the whole Communist era. The modern situation isn't any better.

"These Uzbek businessmen—you never know whether or not they understand a deal up to its end. They prefer a hundred dollars today rather than a thousand tomorrow. And, they even prefer a hundred today rather than a hundred thousand tomorrow. They simply don't believe that this tomorrow does exist."

(Oksana M. Zenina, European Representative, Dunavant Cotton & Ginning Services; conversation with author, May 1993)

"What did I like here in Uzbekistan? For immediate money you can obtain everything immediately. What I didn't like is something that I might not have understood: no deal gives immediate money, so,—for them, it means no deal immediately."

(Vice-President (production) of an American gold mining company; conversation with author, November 1992)

"Don't tell me about Uzbekistan since I know enough. They love bribes, and corruption is the most ancient part of this culture. As one was bought, he would be bought by somebody else later. I can deal with it. What is really difficult is to make them dread bribes as I do . . . I would like to try once more in Uzbekistan."

(Vice-President and CEO of an American oil corporation; conversation with author, April 1993)

"I must repeat this once more: my American colleagues and I met first-class specialists here. But you should know that their activities are almost paralyzed. I would say that the main problem in all the National Republics is this: a real professional has little chance to have true responsibility because he cannot become a boss. Partly, it is a sad result of what they used to call *national politics* as well as a result of patronage and protectionism. You'll find such a system elsewhere around the world. What can possibly stop it? Profit.

In another system, if an enterprise doesn't show a profit, top management changes it immediately. They say in the United States that "money talks." However, here in Uzbekistan, top management is not changed on the basis of competence. Why is this so? Because the salaries of top management aren't dependent on profit which makes wages.

Here, wages — not money — talk."

(Interview given by the author to Uzbekistan State TV,[4] September 1992)

---

4. This excerpt (and some others) was not aired.

Chapter 3
Chechen: Willful Men

◆

Russians would tell you the story that, at the end of the 18th century when the Cossacks came to the eastern foot of the Caucasus mountains to settle there, the arable land (lowland) did not have any owners. Of course, the Chechens would tell you that both this lowland and the highland in the mountains were theirs before the Prophet was born. And both of them have ready proofs.

◆

Rule 1–29
Avoid participation in any discussion about international relationships.

If you join local opinions, they won't believe you; if you raise an objection to it, they will seek to kill you.

Chechens hate to farm. Their women work the fields because it would be disgraceful for a man. In order to obtain control over the region while conquering the Caucasus (first half of the 19th century), the Russian government tried to lure the Chechen nobility out of the mountains (the highland) by giving them large estates on the plain. These lands remained uncultivated for dozens of years.

The Chechens are warriors, shepherds, and traders. They are Shiites, and none of them converted to Christianity during czarist times, unlike a lot of other Caucasians. Chechens believe nothing and nobody except money and himself and a friend from his own tribe; the only law a Chechen would obey is the word of his tribe's elder, with one exception—vendetta.

Due to the national custom of hospitality, a Chechen cannot kill his guest (any man in his house and yard). So a subject of vendetta would come into his persecutor's home and serve as a slave, never crossing the gate. In 1944, when all Chechens (almost 200,000 people) were banished from their republic to Kazakhstan by Stalin, some 17,000 thousand slaves also were expelled separately to Siberia. They say that slavery is still enacted in the modern Chechen Republic.

The city of Grozny — capital of the Chechen Republic —was established as a fortress against the forays by a famous Russian General Yermolov, the subjugator of the Caucasus. His monument in the center of the city has been blown up 13 times and re-established 12 of those times. It doesn't exist now, since the Chechen Republic has declared its independence from the Russian Federation (though Moscow still considers Chechen to be a part of it).

In the beginning of this century, only 0.4% of the population of Grozny were Chechens; before the Second World War, only 8%; after that War, 0%.

Since the Chechens and their neighbors—the Ingush—hate each other (in spite of speaking similar languages), it was logical for Stalin to join the two nations within one Chechen-Ingush Autonomous Republic. Traditionally, the first secretary of the local Communist Party Committee was a Russian, the second secretary, a Chechen; the third secretary, an Ingush. This tradition worked until the German army approached Grozny in 1942 seeking oil.

The Chechen elders declared that the Chechen language is Germanic and their culture is Teutonic (no analogies to Chechen and Ingush languages exist, which is very often the case for most of the Caucasian languages); and they met the Germans like brothers, bringing them gifts, riding white horses (a Muslim habit), and trying to prevent (in vain) the Russians' intention to blow up the oil wells.

The Great Father of all the people—Mr. Stalin—didn't like this one bit. In 1944, half a year after the Soviet Army re-conquered the North Caucasus region, all the Chechens and Ingushs were deported and dispersed to different areas of the central Asian steppes and the Siberian taiga.[1] They were brought back by Nikita Khruschev in the beginning of the 1960s, and they found other people in their former homes in the lowland though their highland villages were empty. They were given money, and they settled themselves in their former homeland—among a diverse population of different nationalities. A few tribes returned to the highland also.

1. Subarctic coniferous forest that starts at the end of the tundra — composed predominately of firs and spruces.

The following population statistics can be found in nationalistic newspapers of the FSU:

- The Chechen population in Grozny during 1953–1960 was 0.02%; 109 homicides—16 per year. (In 1953, after Stalin died, Mr. Beria freed the criminals from the FSU camps and prisons by issuing a general amnesty, and the rate of crime in the FSU rose by 900%.)
- The Chechen population in Grozny from 1960 to 1970 was 11%; 651 homicides—65 per year.
- The Chechen population in Grozny from 1970 to 1980 was 24%; 1,006 homicides—100 per year.
- The Chechen population in Grozny from 1980 to 1990 was 31%; 1,602 homicides—160 per year.

The population of Grozny was never more than 400,000 people. The population of New York City is almost 25 times larger; people murdered in 1993—1,960.[2]

◆

Due to tense national relationships, the political situation around Chechen and within the whole of the North Caucasian region (large, rich, and very attractive for foreign investments) is very confused and complicated. It was in Gorbachev's time (he himself originated from Cossacks of this region) that the KGB suggested to the elders of the lowland's Chechen tribes to contact the Cossack's leaders (newly appointed by KGB, of course, since real Cossacks were annihilated in the FSU long before the Second World War). It was a wise move, which prevented bloodshed in Chechen. Yeltsin's government has chosen another group of Cossacks to send a signal to Chechen, and they understood this signal since no enemies (tanks, jets, paratroopers, etc.) could scare Chechens except the name of Cossacks. Since the empowered Cossacks would not defend the Armenians, Jews, Ingushs, etc., these people got out, probably forever. Those Russians who had a place to go and didn't want to test their luck by staying have left also.

◆

> "My wife died a few years ago. Lucky she was not to see what is going on in her city. Is it possible to work here? Each month I call the president and ask him for the annual salaries for my colleagues and myself. And he answers that all the money is stolen. Is it

2. From a major U.S. newspaper; January 10, 1994.

possible to leave for elsewhere? The very moment I try to sell my apartment here, I will become a 'traitor' of a nation to which I and my sons don't belong. But the nation to which I do belong wouldn't give me the money to buy a place to live among them. We are going to vote in Stavropol—we vote there since it is forbidden in Grozny—and we vote for Zhirinovsky. What other choice have we?"

(Professor E. V. Sokolovsky, Director of the North Caucasian Petroleum Research Institute; conversation with author, Moscow, November 1993)

"We are Muslims, therefore we have plenty of oil. I don't know how much. Professor Sokolovsky will tell you. What I know is that our output per person is the highest in the former Soviet Union. Also, we own our refinery in Grozny. And we have the pipelines northward up to Siberia and southward down to the port of Batum on the Black Sea in Georgia. And our relationship with President Gamsakhurdia of Georgia is excellent, so we are free to transport our oil to the world markets. Probably Mr. Galburaev will detail some technical features. We are rich, we'll prosper, and it is worthwhile for Americans to deal with us."

(Deputy Prime Minister of Chechen Republic; conversation with author, Houston, November 1992)

"Nobody else but the Chechen Foreign Minister had told me that they would fight Russians to the last man. I supposed that it is exactly the thing Russians are dreaming about. . . . And do you know what? This mullah, the North Caucasian's Mufti—I have forgotten his name—he tells me he isn't against a war there since the Caucasian men have become weak. He says, 'A war makes men and people strong.' A pity I didn't have him in Vietnam with me."

(CEO of an American oil corporation; conversation with author, San Antonio, March 1993)

"We are clear, and I repeat once more: we want *national* independence. Since we aren't an empire, we cannot play their games on proportional representation of each nation everywhere in the government, industry, education, and so on. We will do it after we have less than 10% of other people as our compatriots here. Then we may respect minorities. As for today, if other people have to leave, it isn't the matter for me to care. I am not their president. I am the president for Chechens."

(Johar Doudaev, President of Chechen Republic; conversation with author, Houston, November 1992)

These passions along Russia's borders are too strong and important to be easily stifled. Some problems might be solved if Russians and Russia would be quiet and patient. Otherwise, these passions may destroy their willful hosts.

President Gamsakhurdia of Georgia (who spent a few years in a Soviet jail for anti-Soviet and nationalistic activities, then *repented* under KGB pressure in 1984) won 90% of the votes in the first free elections in Georgia in 1991. One year later (as a result of a *coup d'etat* supported by Russia), he was removed by E. Shevardnadze, FSU foreign minister in President Gorbachev's government and former chief of Georgia's KGB and former first secretary of Georgia's Communist Party. Mr. Shevardnadze won 90% of the votes in the second free election in Georgia in 1993.

President Gamsakhurdia ran away to Chechen. Gamsakhurdia, President Doudaev of Chechen, and Russia supported the separatists in Abkhazia (former Autonomous Republic of Georgia). And, after a bloody war, Abkhazia won independence from Georgia. Then President Gamsakhurdia, with his (and Chechen's), troops tried to reconquer Georgia's presidency for himself but failed (since Russia now supported Mr. Shevardnadze). He returned to Chechen and committed suicide (or was killed) in Grozny on the first day of 1994.

The U.S. position and participation in these events are unclear. Mr. Shevardnadze had and has excellent relations and connections with the U.S. administration and obtained a few instructors for training special troops. Mr. Ames, CIA officer and Russian spy, visited Georgia in 1993 and one of the instructors was shot to death near Tbilisi (Georgia's capital) soon after Mr. Ames left. This might be the first sign that Russia was unhappy with the American appearance within Russia's sphere of influence. And, of course, after a certain time, Russia will try to include Abkhazia, with its sunny resorts and important Black Sea ports, into the federation.

> "No, you cannot compare these issues. Georgia is a country of 5.5 million people and they wanted, fought for, and won independence. First, 90 thousand Abkhasians or 50 thousand Ossetians[3] just cannot survive independently. Second, their ancestors were invited here to guard our land[4] and still they are minorities in those districts where they live. Third, Russia would have swallowed them immediately if they had become independent. We don't want one piece of our land to belong to Russia; we don't want anything in common with Russia. Now, you asked about Chechen—they are stupid and therefore brave. Of course, a few hundred thousand Chechens mean nothing for Russia, but if they were to guard Georgia's boundary, we would support them. Gamsakhurdia tried but couldn't win enough time. . . . This and only this is their prospect for the future, believe me."
>
> (A. Babilashvily, Professor of History, University of Tbilisi; conversation with the author, Istanbul, March, 1993)

> "Come, let us go down and confuse their language there, so that they cannot understand one another."
>
> (Genesis 11:7)

◆

The five thousand year history of man has caused some men to declare that God's will has worked in mysterious ways. You could even make a claim that the rise and fall of emperors and empires reflect man's disobedience, since these nations tried to force a kind of peaceful co-existence between different peoples. But each national culture has sooner or later resisted any forced attempt at assimilation. The national cultures' response to these attempts is often disproportionately irrational and cruel (Balkans, Caucasus are the latest

3. South Ossetia was an Autonomous Republic of Georgia while Northern Ossetia was an Autonomous Republic of Russia (Stalin's joke!).
4. Late in the 12th century, Georgian Princess Tamara used foreign troops to guard boundaries of her Tsardom.

examples) since the empires' initial impulses and requests, from the very beginning, have been unjust and disproportionately outrageous. The subjugated peoples are doomed to find the subjugator's culture disgusting and repulsive, and those personal carriers of that culture — outlaw.

One of Ivan the Terrible's wives was a Caucasian princess. During the conquest of the Caucasus, and after, Russian high society was open for Caucasian princes; Russian universities welcomed Caucasian students; Russian brides welcomed Caucasian grooms. Yet the *assimilated* Caucasian people participated in destroying the Russian Empire willingly, readily, and gladly.

Primarily, the Communists honestly tried to find another way to rule their multinational country by declaring and organizing a status of national autonomy for minorities.

◆

> "It would be unforgivable opportunism if . . . we should undermine our prestige . . . with even the slightest rudeness or injustice to our own minorities."
>
> (V. Lenin, 1922)

Under Stalin's rule (he was a Caucasian himself), the national policy degenerated into routine Russianization, which was disguised with emphasized slogans of equal opportunities for non-Russian minorities. In order to make this great show more convincing, it started to be introduced first in the area of education since almost 75% of the FSU's population was illiterate in 1929.

And until 1991, each Russian university, art school, diplomatic school, medicine and technological institution, economical and financial school, trade and business institution, Communist party high school of politics, etc., were obliged to have reservations for students from the Unions and Autonomous National Republics at their request.

Result 1: Each (without any exception) president, prime-minister, deputy prime-minister and almost each minister and deputy minister and member of the local Parliament of each newly independent state (former republic of the FSU) graduated from a Russian high school. (President Djohar Dudaev of the Chechen Republic graduated from a high military school and then — due to *equal opportunities policy* — became a general of the Soviet Air Forces.)[5]

---

5. The gossips insist that he`stole a few nuclear bombs and brought them to Chechen. They allege that this is why the Russian government speaks so softly and nicely with the Chechen government.

Result 2: Since the local authorities chose the individuals to be sent to Moscow (to Russia), the most able persons were never chosen, but the relatives of the elite or those who had bribed the senders became the students (more than 90% of them represented the national minorities).

Result 3: Since Russian high schools were obliged to have a certain percent of the national minorities among their graduating students, they would issue diplomas to each minority student who entered the school even though he (she) remained highly ignorant.

Result 4: Returning home, these specialists took their position in industry (art, trade, medicine, etc.) according to their connections with the local elite (or the money their parents had) and began to make important decisions for industry. Their incompetence destroyed the FSU's economy.

Result 5: No qualified specialists remained in the national republics, since the people of other nationalities left.

During the 1970s and early 1980s, many of Moscow's (Leningrad's, etc.) Jews lost their jobs because they applied to emigrate to Israel and were refused. Some of them, being specialists in different areas, made money for a living by preparing Masters and Doctors dissertations for individuals representing the national minorities.

From 1976 to 1984, one such team of specialists prepared:

- 8 Ph.D. dissertations in medicine for individuals from Azerbayzjan, Moldavia, Chechen-Ingush Republic (it was one republic until 1991), Georgia, Kirgizstan, and Turkmenistan
- 5 Ph.D. dissertations in chemistry for individuals from Uzbekistan, Georgia, Kazakhstan, and Uzbekistan
- 5 Ph.D. dissertations in physics for individuals from Abkhazia, Georgia, Uzbekistan, and Turkmenistan
- 4 Ph.D. dissertations in petroleum geology and drilling for individuals from Kazakhstan, Turkmenistan, Uzbekistan, Chechen-Ingush, and Kalmyk
- 19 Ph.D. dissertations in philosophy, linguistics, sociology, and history

Forty-one dissertations were made, and 32 individuals received Ph.D. degrees; 9 individuals didn't know what they were talking about while giving the oral report to the Scientific Council, so they missed the opportunity for a degree.

Neither this team nor these Caucasians invented such a business. It was introduced by the FSU establishment. A man who didn't miss the opportunity was Mr. Baybakov (who originated from the Caucasus region), the oil minister in N. Khruschev's government and the chairman of Gosplan (State Planning Committee) and deputy prime minister in Brezhnev's time. Obviously, this gentleman—in order to enhance his prestige and to be prepared for any surprises in the future—decided to become a Ph.D. and a professor, since the wages in these positions were the highest in the country. It must be noted that the elite professors of IGIRGI (Institute of Combustible Fossils) promoted this idea to him and prepared his dissertation, since he was a very busy man, of course. In the beginning of the summer of 1975, the dissertation was brought to the Scientific Council of the same IGIRGI. Yes, the same professors who prepared the dissertation had to decide whether or not it was worth a Ph.D. degree.

Being very busy, Mr. Baybakov was represented by . . . his wife, a teacher of Russian literature. After the Council made a positive decision, she said that she would let her husband know that this Council showed friendship to her husband.

Rule 1–30

While traveling to the republics outside of Russia, never visit a physician who is a native of that republic.

The chances that a physician of a local nationality bought his diploma are huge. Seek a doctor who is Russian, Jewish, or Armenian; they have never participated in the equal opportunity game.

Rule 1–31

Avoid appointing a specialist of a local nationality to a position which requires knowledge and responsibility in your (joint) venture.

Especially avoid people who graduated from the Moscow or other Russian high schools since they were involved in the equal opportunity game. Those, nevertheless, might be useful for keeping warm relations with the local establishment.

Rule 1–32

If your partner's background includes education outside of the local region, you should be aware that he doesn't understand the deal you are negotiating, since he is likely to be ignorant in the technical and financial details of the negotiations.

This individual would never tell you that he doesn't understand something; he would try to avoid any details which might be important to you. He wouldn't like you to insist on clarifying those details, and the negotiations will collapse. Try to have his deputy and some engineers at the table.

Individuals who graduated from the local high schools are more reliable since they have to study the subject in order to obtain their diplomas. Some more statistics are given below.

About 9,000 students graduated from the Grozny Petroleum Institute (old, famous, respectable) during 1980–1990. They obtained specialties as petroleum geologists, geophysicists, field engineers, refinery engineers, drilling engineers, etc. Only 11 Chechens were among them, and these specialists reached high positions in the oil industry but not in their republic because they lacked proper connections (or money). You can assume that these persons were interested in their chosen fields. At the same time, more than 90 Chechen students graduated from the Moscow Petroleum Institute. About 60 of them took high positions inside the Chechen Republic, and only 7 of these positions were in the oil industry.

> "Do you know why I haven't any good specialists here in Chechen? This is very simple. There was almost nothing to steal in the oil industry. Now it has suddenly become a very profitable deal. So those individuals who graduated from the Moscow Petroleum Institute and who were working in trade or warehouses, or stealing bricks while heading construction works, remind themselves that they are oilmen. And they try to remind me of this. I resist, but they will kill me soon."
>
> (Ruslan Galburaev, Minister[6] of the oil industry of the Chechen Republic; conversation with author, Houston: November 1992)

---

6. Mr. Galburaev, former chief engineer of *Sakhalinneft* (Sakhalin oil) was invited by President Doudaev to Checken and was deputy minister of the oil industry until the president dispersed the parliament; then Mr. Galburaev became the minister. Mr. Galburaev was removed from his position early in 1994.

"President Kennedy and President Johnson didn't invent anything new by introducing affirmative action programs, since the FSU was first here also. I wish they would have gone there to study the results before pushing these programs in the United States . . . unless Mr. Johnson really wanted the United States to reach the level of incompetence I have seen everywhere in Kirgizstan or Dagestan or Chechen-Ingush, etc., while working there."

(Peter Buckly, Vice-President, Petron Industries Inc.; conversation with author, Houston, March 1993)

For able pupils, no educational system can become an obstacle in obtaining knowledge. There are plenty of highly educated people and first-class specialists in the FSU. The previously described issues may help to find them, to understand their background, and to establish fruitful relations.

◆

Regarding oil exploration, the centralized system and the state's total ownership had some unquestionable advantages.

In the late 1970s and early 1980s, in almost each basin—except West Siberia, where the incremental produced oil volume also became smaller—oil production was decreasing. The Oil Ministry together with the Ministry of Geology had invested money and set up a program for the complex study and exploration of the major petroleum basins of the FSU. The location of a few dozen regional seismic surveys was selected. These seismic surveys, made by local operating geophysical companies, crossed the largest oil (gas) fields and unexplored areas within each basin in different directions.

◆

In the late 1980s, this program was completed and the results were as follows:

- North Caucasian Basin:
  Eastern Region (Grozny): 8 regional seismic surveys, total length—almost 1,500 line km
  Central Region (Stavropol): 4 regional seismic surveys, total length—almost 950 line km
  Western Region (Krasnodar): 8 regional seismic surveys, total length—almost 2,100 line km
- South Caspian Basin (Azerbaizjan):
  Onshore: 3 regional seismic surveys, total length—almost 500 line km

Caspian Sea offshore: 9 regional seismic surveys, total length—almost 5,400 line km

- Mangyshlack Basin: 3 regional seismic surveys, total length—almost 1,350 line km
- Pre-Caspian Basin: 2 regional seismic surveys, total length—almost 370 line km
- Volga-Ural Basin: 5 regional seismic surveys, total length—almost 1,190 line km
- Pripyat Trough (Byelorus): 2 regional seismic surveys, total length—almost 430 line km
- West Siberian Basin: 2 regional seismic surveys, total length—almost 850 line km
- Sakhalin Basin: 3 regional seismic surveys, total length—almost 670 line km
- Barents Sea offshore: 3 regional seismic surveys, total length—almost 2,040 line km

"I was really impressed. I'd say I envy you having these regional lines. Too bad we cannot afford such work onshore in Texas."

(K.P. McNeill, Vice-President, MD SEIS, INC.; conversation with author, Moscow, January 1990)

"The idea of regional seismic surveys is old. What is new is that you have so many of them. These are good tests to evaluate whether or not geologists have mastered geology."

(Professor Bally, Rice University; conversation with author, Houston, January 1992)

Regional seismic surveys were made across each of the oil basins of the FSU. Ask your local partners to provide this information.

The local operating company of Chechen (*Grozneftegeofizika*) started this program and several new oil fields within non-anticline traps were discovered.

◆

Chechen was and is rich with oil; from 1955 to 1965 the production rate from 5 oil fields rose from 2.1 million tons to 9 million tons and reached its peak at more than 9 million tons per year (the highest rate) during 1965 to 1967. But by 1990, the production rate from almost 20 oil fields declined to 3.3 million tons per year and in 1991, down to 2.1 million tons per year. It is hard to give the production rates for the years 1992 to 1994 since information does not exist. Fifteen years ago, more than 50% of the oil was produced from a depth of less than 1.0 km; in 1991, less than 25% of the oil was produced from these depths.

◆

"If it is secret, it is a secret for myself."

(R. Galburaev, Minister of Oil Industry of Chechen; conversation with author)

And he added a few words which are unprintable. He was neither a nationalist nor General Dudaev's man; he is just an oilman who works hard trying to fight chaos.

More than 950 shallow wells were abandoned since the oil fields are depleted, and more than 100 deep wells (3.0–5.5 km) were abandoned due to various technical reasons.

The study shows that no more than 20% of the oil reserves was produced, even though oil production in the Chechen region began in the year 1821 (58,000 tons of oil were produced during the first 70 years). Less than 35% of the shallow pools' reserves and less than 20% of the deep pools' reserves was produced in total. More than 350 million tons of oil (in-place) remains within the discovered (deep) oil fields and twice as much of the oil resources hasn't been discovered yet. The southern edge of the highland is badly explored because of the sharp relief, which creates problems for the seismic survey. Existing seismic lines show the presence of a large (12 x 7 km) Paleozoic (probably, Triassic) reef as deep as 6.6 km and a few large (10 x 4) folds within the Jurassic and Cretaceous deposits at depths of 5.2 to 6.5 km. These stratigraphical sequences, together with Paleogenic and Lower Neogenic sediments, are oil saturated within the lowland. In addition, during the last 10 years, a few large oil-gas-condensate fields were discovered within Upper Cretaceous limestones along the southern edge of the highland. In order to develop these pools, more than 150 km of reliable roads (and pipelines and other logistics) should be constructed across the mountains, and about 40 deep (5–6 km) wells have to be drilled.

◆

The following dialog is from a meeting which occurred in November 1992 at Houston, Texas.

The Chechen delegation consisted of the Deputy Prime Minister (D.P.M.), the Oil Minister (O.M.), and the Director of the North Caucasian Petroleum Research Institute (D.). The American representatives included the Chief Executive Officers of various American oil and service companies (CEO).

◆

D.P.M.:. As you can see, I am here to negotiate the contract and prepare all of the documents. We are scheduled to sign this contract in the presence of our president, when he comes in a few days.

CEO 1.: You are going to pay crude oil and *mazut* for our equipment and services, aren't you? Do you have any guarantees of delivering these to the port terminals and FOB, since your relationship with Russia isn't very clear or peaceful?

DPM: But then our relationship with independent Georgia is excellent. President Gumsakhurdia of Georgia[7] needs money badly, and he is our ally. We pay currency for transportation of the oil through Georgia to their port of Batum. And I want to assure you—this is a secret—that this, what they call *Chechen mafia*, controls the port of Novorossisk at the Black Sea also. Don't worry.

CEO 2.: Then we can ship our equipment as soon as it is paid for.

CEO 3.: We should insist our specialists use our equipment on your fields, and they will make the interpretation of the data also.

D.: Our specialists are familiar with your devices and the interpretation methods.

CEO 2.: No, they are not.

They are familiar with the equipment and interpretation methods.

O.M.: What we need is one or two American specialists to observe and to help if something goes wrong.

CEO 2.: We will bring the drilling teams, geophysicists, and other engineers.

---

7. The story of President Gumsakhurdia of Georgia has been mentioned previously. After he was removed from office, Chechen supplies of fuel to Georgia stopped. Then Shevardnadze's government stopped transport of Chechen oil through the Georgian port of Batumy.

D.: It isn't the matter we have discussed. . . . We have more experience dealing with our oil fields—decades of experience—and we know better what has to be done there. All we want is the equipment.

CEO 4.: I am sure our specialists must begin first.

It would be better if he added: and teach your personnel.

CEO 5.: Special skills and practice is needed to use our devices. I doubt your personnel is trained enough.

D.: Please, don't underestimate our personnel. We had experience using Schlumberger's equipment at our wells. I want to tell you that the common practice is: when foreigners come to the country, they use the local specialists as much as possible. And it will be cheaper, also. The project will be cheaper; we'll spend less money; you will spend less money.

CEO 6.: Maybe it is true, Professor Sokolovsky. But sometimes the less money you spend the less you would earn. Are you sure that your specialists are qualified to prepare all the data in the computer formats we need?

D.: Tell us your requirements, and we will try.

CEO 7.: Wrong, Mr. Sokolovsky. I happen to know that you have neither such computers nor the software of this kind.

O.M.: Well, we don't have these features. Why do we need all of these? You will bring a rig to workover these 17 shallow wells on an old oilfield; one rig for drilling new wells as deep as 4.0 km; one rig for workover and drilling as deep as 6.0 km. If you want to bring some advanced monitoring systems, Schlumberger's well logging, etc., we appreciate this. As I have understood, a great show is on its way and we appreciate it, also. You want to advertise the possibilities of the modern technology—you are welcome; but you pay. We will pay for the rigs we need and for the three drilling teams, as I have said. And we want new rigs!

CEO 2.: New rigs don't exist. You can order them from IRA (a U.S. company) and they will cost you three times more than my rigs. And my remanufactured rigs are like new. You will see them before you buy anything.

He should add: I'll get IRA's prices on their new rigs, and we'll compare those to my prices.

CEO 8.: I am excited, gentlemen. Why? My company is going back to Russia and to exactly the same place where it started at the end of the last century—to Nobel's oil fields in Grozny.

It is true, but he got in a "dig" at the Chechens' feelings by using these words "back to Russia."

CEO 8.: . . . And since we are all specialists, you do understand that drilling costs a lot but brings nothing if the hole isn't explored. The very purpose of drilling a hole is to obtain information as to whether or not there is oil there—and where it is. Believe me, Mr. Galburaev, and you, Professor Sokolovsky, that without this *show* we bring to Grozny, not one drop of oil will show on the surface. It may have happened long ago, but you cannot rely on this now. I am very familiar with how the Russians make well logging, and it is worth nothing.

He made it unknowingly, but it was an essential ploy. Most oilmen in Chechen, including the geophysicists, are still Russians. The geophysical equipment rather than the specialists has to be blamed for poor well logging results. Nevertheless, the last statement added something to the deputy prime minister's opinion, which, probably, might be like this: the smaller the number of Russians involved in the deal, the better.

D.P.M.: I would rather join the American team at this point. And since I am here to make decisions, we are going to pay the full price of all services. Let us discuss this price.

A graphic example of how political prejudices overcome economical expediency. Of course, it isn't his own money which will be spent.

CEO 2.: I know you have a few regional seismic surveys crossing the Chechen Republic and Dagestan. We'd like to have a look at them.
O.M.: We are ready to sell this data.
CEO 2.: I don't like it. I want it for free.
O.M.: I want your equipment for free.
CEO 2.: You'll have a free look at my equipment, and I'll have a free look at your lines.

Rule 1–33
There is a long distance to go between signing a contract, receiving money, and initiating activities on these projects. Be patient.

◆

A contract means almost nothing in this part of the world since a system of laws and courts, which would make the agreements binding, doesn't exist. A contract expresses a balance of interests of each of the partners. While the state was strong, its representatives had to follow the state's interests. Since the state became weak elsewhere within the FSU, its representatives are looking for their own profit in each deal.

Russian Prime Minister V. Chernomyrdin denounced a contract with Conoco on exploration and development of the offshore area of the Barents Sea (May 1993). Russia's military promised him to invent a technique of exploration drilling from submarines; of course, he preferred to keep them busy rather than the Americans.

The story of Chevron's denounced contract on Tengiz was the first signal of President Gorbachev's amazing weakness (January 1990).

The offshore Sakhalin area (Russian Far East) has changed contractors three times. The reason: rivalry among different local and Moscow authorities (1990–1992).

The Parliament of Kirgizstan almost declared *null and void* the contract on the Kum-Tor gold deposit, signed by the prime minister and approved by the president (November 1993).

◆

Rule 1–34
It is very easy to violate any agreement anywhere in the FSU.

This rule means that it is possible for foreign partners to play such games and to escape with minor troubles. Don't misuse this opportunity.

> "The first check signed by President Doudaev was given to his aide, who had to leave for England next day. This check was never deposited in the bank; both the aide and the check disappeared forever[8] (so

8. In 1993 a few Chechen men were killed in their own houses at fancy blocks in London, Istanbul, Nikosia, and Berlin.

I was told). The second check was handed directly to one of our partners, who made a special trip to Grozny. When I came especially to take the third check, the president tells me that the signed check is in the finance minister's office. I went there with my partner, a former U.S. Marine. This minister tells me that the check was stolen from his office yesterday. I wasn't amazed. I just invited this minister close to a window. Embracing his waist, I explained to him that I was going to throw him out of the window. It was on the fourth floor, and he understood that he wouldn't survive. So he gave the key to my friend and showed him the place—just beneath President Doudaev's full-length portrait. It wasn't so easy to leave Grozny the same day."

(CEO of an American corporation; conversation with author, Moscow, December 1993)

A short excerpt from a conversation between the author (A.); Mr. Rouslan Khasbulatov (R.K.), the Speaker of the first Russian Parliament; and Mr. Vitaly Urazjtcev (V.U.), member of the first Russian Parliament follows (White House, Moscow, June 1993):

A.: Rouslan Imranovich,[9] please, what good advice would you give to those Americans who work with your fellow countrymen in Chechen.

R.K.: A clever American?

V.U.: As I have said to my friend Roy Romer, the governor of Colorado: no clever American would wish to work in Russia.

A.: Would you sign this statement, Rouslan Imranovich?

R.K.: No comment, since he knows Russia better. I would sign another: be assured that my fellow countrymen will outwit . . .

A.: Let me put it like this: no honest business between unequal partners is possible. Until the Chechens are doing business in the USA, the Americans are out of business in Chechen (in Russia, etc.). Is this what you try to tell me?

9. The Russian way to address a person is to use his name and his father's name; Imranovich is the special form of Mr. Khasbulatov's father's name.

R.K.: You are using some moral terms in a wrong way. We are equal.

V.U.: Even more equal than others.

R.K.: Yes, as we know. But our economical possibilities are quite different. So we must use other rules and other games.

V.U.: So keep yourself safe, Joseph.

A.: Am I menaced with something?

V.U.: Between these two worlds one becomes crazy very easily.

Mr. Khasbulatov, Chechen by origin, graduated from a Moscow high school and became a Ph.D. and a professor of political economy. He was the first speaker of the first Russian Parliament. He was arrested in October 1993 after this Parliament was dissolved by President Yeltsin's troops.

Mr. Urazjtsev was the chief of the Parliament's committee on reforming the Army. He defended the Russian White House in October 1993; he was arrested and then released.

In February 1994, the newly elected Russian Parliament granted amnesty to all those enemies of President Yeltsin.

◆

The other North Caucasian Republics (Dagestan, Adygey, Ingush, etc.) aren't independent either politically or economically. Money, supplies, equipment, etc., for their industries (oil industry) come from Russia, and Moscow is in control of their foreign activities.

Some time ago, a group of petroleum companies (one British and two American) signed an agreement for onshore and offshore exploration and production with *Dagneft* (Dagestan oil). Mr. Lisovsky, the former head of the Geological Management of the former Oil Ministry of the FSU, quite occasionally got to know much that was new to him. He and his office weren't too busy at that time. So he found a job for himself, writing a letter to the chief of *Dagneft* to warn him about the negative consequences of this *insufficiently considered cooperation*. Mr. Sayidov, the chief of *Dagneft* was brave enough to show this letter to his foreign partners, but the deal was postponed. Mr. Lisovsky's (L.) point of view was expressed in a conversation (Houston, Texas; February 1993) with the author (A.) which follows:

◆

A.: As I know, you haven't read this agreement between *Dagneft* and the foreign companies. Is it true that Mr. Jabrailov, the chief geologist of Dagneft, just bragged of his success in this deal?

L.: He told me all the details I needed.

A.: Were the terms of this agreement really insufficient?

L.: Yes, they were. *Dagneft* should have insisted on the investments and exploration being made in a limited period of time. (The time frames weren't pointed out at all.) Why did they give the foreigners 55% of the increased oil production? And they wanted 85% of the increased production in order to more quickly return their investments. It was ridiculous. These people from *Dagneft* have no experience dealing with foreigners. So stay away from them.

A.: Did you count the geological risk, the political risk, the investments, and the reward?

L.: You are crazy! Why should I?

A.: You could have proposed making a very qualified conclusion after the study of this project, since you had the best personnel for that in your Moscow office.

L.: What for?

A.: First, you would improve the terms of this agreement for *Dagneft;* second, you and your personnel would earn some money being a mediator. Both sides would pay you willingly and gladly for the expertise.

L.: I don't understand what you are talking about.

A.: You have given free advice to *Dagneft*. You could have charged money for this qualified advice, couldn't you?

L.: Don't call this advice. I would call them orders. Each order is reasonable since it is obeyed.

> "They have at least a thousand Mercedes in Chechen, but the roads are suited only for Jeeps. Guess why? Money paid for their oil is stolen. Russia delivers three times more oil to their refinery; this money is stolen also. There is no money for road construction. Some of our people prefer Russian control since not *all* the money is stolen. Compare the roads, for example."
>
> (Magomed Jabrailov, Chief Geologist, *Dagneft;* conversation with author, Moscow, June 1993)

> "It seems the people there don't know what they want. And I am not the person who would tell them this, since I don't know what I want there either."

(Vice-President of a large American oil company; conversation with author, Minsk, August 1992)

"Americans deceived us, at the last moment. The rigs they brought are used and old, covered by mud and rust. I am not sure it is possible to use them here. And they aren't sure, also, since they claim now that it is dangerous for their teams to work at our oil fields. It is a ploy."

(Rouslan Galburaev, Oil Minister of the Chechen Republic; conversation with author, Moscow, November 1993)

No, it wasn't a ploy. In the beginning of 1994 the company sent a team to prepare the rigs for drilling, even though the area really wasn't safe.

◆

In the middle of the 1970s, I happened to be in Grozny together with the retinue of the deputy minister of the Oil Industry of the FSU, Mr. Orujev (originated from Caucasus). A noisy feast honoring him ended at 3:30 A.M. and drunk men were leaving the restaurant for their cars. One of them embraced my shoulders, looked straight into my eyes, and said pretending to be dead drunk, "I have deceived everybody. I have paid the bill. . . ."

◆

Of course he wanted me to tell Mr. Orujev that it was him who paid the bill.

Early in 1993 I met this old man[10] in Moscow. They say he is a VIP of the Chechen international mafia after a very successful career in the former Communist Party.

"It seems that your people keep deceiving everybody—paying that money which someone else has spent—doesn't it?" I asked.

"You asked about my people; they know why they pay. We are willful and free. The first is forever; the second—a short while.

These people need willful men to work with them. Some Americans are. . .

---

10. Permission was not granted to use this man's name.

Chapter 4

Tatarstan: Old Energy

◆

The Tatars ruled Russia for a long 300 years. Then Russia ruled the Tatars for another 300 years. Russian Princes served Tatars Khans. Tatars Khans served Russians Tsars. Usually after capturing a Russian fortress, Tatars decapitated each male taller than the tongue of a cart. Ivan the Terrible conquered the fortress of Kazan (on the Volga River; now the capital of the Tatar Republic; Fig.1–1) and ordered the throats cut of each living thing, including horses, dogs, and cats. After a while, he abdicated and appointed a Tatar Khan as the Tsar of Russia in Moscow. Among their other titles (the Tsar of Poland, the Great Prince of Finland, etc.), Russian Emperors carried the title: the Great Khan of Tatars. Russian troops (Orthodox Christians) helped Tatars (Moslems) fight Lithuanians and Poles (Catholic). The Tatars were excellent soldiers in the Russian Army during the World War I and the real Heroes of the Soviet Army during World War II.

Those inhabitants of the Autonomous Republic of Crimean Tatars, which hadn't been in the FSU Army, met the Germans gladly, like their best friends, in Crimea in 1942. Comrade Stalin deported the whole Tatar population (including war-disabled Heroes of his Army) from Crimea (they lived there for 700 years) to Siberia and central Asia.

Not to Tatarstan.

Here are some statistics from Moscow's daily *Pravda*[1] (May 8, 1975) concerning the Heroes of the Soviet Union (the FSU's highest military decoration).

- Tatars: 964 (total population before the war—about 2.0 million people)
- Jews: 1,198 (total population before the war—about 4.0 million people)

1. *Pravda* (The Truth) was the newspaper of the Central Committee of the Communist Party of the FSU. It remains the communist daily newspaper.

- Byelorus: 1,898 (total population before the war—about 7.0 million people)
- Ukrainians:3,717 (total population before the war—about 35.0 million people)
- Russians 9,188 (total population before the war—about 75.0 million people).[2]

It seems during this war, Comrade Stalin was unhappy with the behavior of most of the nationalities living along the western and southern edges of his country. (Byelorussia was the only exception since a quarter of its population had been lost during this war.) And the punishments for them were fast and cruel. After the war, more than 1.5 million people were killed or deported from the Baltic Republic (total population before the war — about 5 million); more than 0.4 million people were deported from Moldova (total population before the war — about 2 million); all the Tatars were deported from the Crimea; all the Moslem nations of the North Caucasus were deported; Comrade L. Kaganovich (a Jew) was appointed as the first secretary in Kiev, Ukraine (an essential ploy to show Stalin's distrust and to disgrace the Ukrainians); and, a few Tatars were appointed as the rulers of some western regions of Russia itself (to remind Russians of the Tatar's yoke).[3]

Tatars and Russians — the two nations have a long history of living together; expending energy to be enemies has become very old.

> "Each Russian carries 25% or more Tatar blood in his veins; therefore, we aren't enemies; we are cousins, if not brothers."
>
> (A. Ostashvily[4], Vice-Chairman of the Russian National-Patriotic Front *Pamyat*; said at a meeting in Moscow; June 1988)

Some 51% of the population of modern Tatarstan are Tatars; some 49% of the population of the capital city of Kazan are Tatars (almost one million total population), and no other city's population has more than 33% Tatars. These demographics together with all

2. These figures of the total population weren't printed in *Pravda*.
3. The 300-year period of Tatar rule over Russia is called *the Tatars' yoke*.
4. He was arrested after a pogrom in Moscow's writers club, sentenced to two years in jail, and supposedly committed suicide in a prison in the city of Vladimir in 1989.

evident geopolitical factors—Tatarstan is located in the middle of Russia (Fig. 1–1)—leave the Tatars' nationalistic and separatistic movement without any hopes, even though the movement is underway. (Tatarstan, for example, didn't participate in the elections for the Russian Parliament in December 1993.) Of course, this kind of an economical autonomy, which Tatarstan enjoyed during early 1990s, might be expanded.

> " . . . and take as much sovereignty as you can eat up."
>
> (B. Yeltsin, Chairman of the Supreme Soviet of the Russian Federation; said at a meeting[5] in Kazan, May 1990)

◆

Tatarstan's oil (together with Texas oil) helped to win World War II. Tatarstan's industries were major participants in this effort. Today these include the chemical industry and many manufacturing industries (aircraft, automotive, rockets, submarines, lasers, instruments, machine tools, etc.).

The peak of Tatarstan's oil industry was reached in the late 1950s and early 1960s, when almost 100 million metric tons of oil per year were produced. Since then, the production rates declined to less than 15 million tons in 1993.

◆

There are no refineries in Tatarstan.

The secondary and tertiary oil resources within Tatarstan's major oil field might be as much as 600 million metric tons; the estimated oil resources within the promising areas—less than 300 million tons (in-place). All these oil resources are trapped within the limestones and sandstones of Devonian and Carboniferous deposits as deep as 1.2–2.2 km. These deposits are covered by Permian and Triassic sequences. Said sequences contain enormously huge bodies of solid bitumen. In some areas of Tatarstan, the dimensions of such bodies are as large as 56 km in length, 13 km in width and 490 m in thickness. Probably, within the Triassic Period, some of the huge hermetically sealed oil pulls had become depressurized,

5. Mr. Yeltsin wouldn't repeat this sentence after he became the Russian President.

possibly by erosion, and the crude oil seeped to the surface and formed large lakes in the deep depressions. The described bitumen contains up to 6.6% sulfur. (Local oil contains up to 3.6% sulfur.)

> "Our government is looking for somebody who knows how to utilize this bitumen. We will accept his terms and clauses in advance."
>
> (A. Rijanov, Chief gas and oil expert of the Cabinet of Ministers of Tatarstan; conversation with author, Kazan, April 1993)

> "I spoke with the Tatarstan prime minister by phone. I told him that we have a company which could bring a cracker to refine their bitumen. He was excited; he arranged a meeting with me in Moscow this week. He was in Moscow; this week is over. He left for Kazan. No meeting took place. I am excited."
>
> (Orhan Sadic-Khan, Managing Director; Paine Webber Inc.; conversation with author, Moscow, November 1993)

Rule 1–35
Don't expect punctuality from the local VIPs.

An Oriental VIP must perform willfully. But you follow this French proverb: "La ponctualite est la politesse du Roi." *(Translation: The punctuality is the politeness of a king.) Be a king. In time they will respect you. You have nothing to lose by following this proverb.*

◆

During the era of the FSU, Tatarstan's government (like each of the local governments) had nothing to do with the local oil and gas industry. All the oil enterprises in Tatarstan reported directly to Moscow and obeyed orders coming from the Oil Ministry in Moscow only. Since 1992, this kind of relationship changed drastically in Kirgizstan, Uzbekistan, and the Chechen Republic but changed very slightly in Tatarstan. An unpublished agreement signed by Russian President Yeltsin and Tatarstan's President Shaymiev stated an increase in Tatarstan's share in the production of local industries (including oil production). And, the Tatarstan government would have total ownership of any future enterprises it created.

The first clause is continuously sabotaged by the heads of the large local enterprises. For example, a rich Russian bank wanted to buy 10 Kamaz trucks from a plant in the city Naberezjnie Chelny (formerly named Brezhnev). The bank's representatives (B.R.) showed the Sales Director (S.D.) of this plant a multiple purchase order signed by the Tatarstan prime minister. The following dialogue took place:

◆

S.D.: This guy wants to be paid for nothing.
B.R.: But these 10 trucks—they are from the Tatar's quota this year.
S.D.: No way. We don't have any trucks for sale this year.
B.R.: Well, next year . . .
S.D.: I can promise nothing.
B.R.: I have a multiple purchase order from Moscow.
S.D.: Please use it.
B.R.: Now?
S.D.: Why not? Call your office and let them pay the money for the trucks in Moscow.

A fee was promised to these representatives by Tatarstan's government if they would buy these trucks out of the quota and pay money in Tatarstan.

The second clause in the agreement remains theoretical since the Tatarstan government doesn't have money to create new enterprises.

Rule 1–36
When starting a business, try to keep the appropriate Moscow authorities at least neutral.

It doesn't mean that you need to inform them in advance; it means that the source of their information must to be chosen by yourself. Your partners may decide to report to Moscow in an especially negative manner; prevent this from happening.

The idea of being introduced is very important everywhere and is a matter of vital importance within eastern countries. The following two examples illustrate this point.

A delegation from a large and respectable American engineering company came to the FSU early in 1993. After the negotiations, their chief, who was sure that his company was known everywhere around the whole world (including Tatarstan), decided that Tatarstan should be included in this trip. He called from Moscow to the office of Tatarstan's prime minister and tried to make an appointment . An assistant of the deputy prime minister promised to set up a meeting with his boss. In Kazan, nobody met the delegation at the railway station, and they went to the building of the Cabinet of Ministers by themselves. Not only the deputy, but the same assistant sharply refused to see these Americans. They were lucky they could get tickets back to Moscow the same day.

The president of a small American oil company had sent a letter directly to President Shaymiev of Tatarstan, advertising his company for secondary recovery at the Romashkinskoe oil field (the largest in Tatarstan). After a few months, no response had been obtained. So, the vice-president of this company together with a geologist were sent to Kazan. From Moscow, they called Mr.Rijanov (mentioned above), whose phone number is listed in a reference book (edited in the United States) of the FSU's oil industry VIPs. Mr. Rijanov was very polite, since he was the one who was supposed to have answered the letter, but he hadn't bothered to make the reply in time. So he took the time and had a conversation with these two Americans. He signed a protocol with them and then — a month later — sent them a telex: "We are sorry to inform you that another company has been chosen for secondary recovery at Romashkino." Before he met these two Americans, he knew that a contract with another company had been approved by President Shaymiev.

Rule 1–37

Be introduced to local partners by a Moscow VIP. Come to visit your local partners *after* you are invited by them.

The chain (of people) may be long until you reach the final person who will introduce you; it doesn't matter. Reference letters may be useful; send them first to obtain in writing an invitation to visit the country. This is your local agent's duty also. If recommended from Moscow, search out the nationalities of both persons: the person introducing you and the person to whom you are recommended. It is better if the latter is of the local nationality.

◆

An American oil company wanted to construct oil refineries in Tatarstan. In April 1993, in the city of Kazan, a meeting was held with Mr. Albert Rijanov (A.R.), chief gas and oil expert of the Cabinet of Ministers of Tatarstan and Mr. Marsel A. Dautov (M.D.), Minister of Community Services of Tatarstan. The American representatives were the vice-president (V.P.) of an oil company, the chief engineer (C.E.) of an engineering company, and the author (A.)

◆

M.D.: Thank you, Mr. Vice-President. I am very impressed with the history, size, and financial possibilities of your company. Now I can recall in my mind the name of your company, Mr. chief engineer. I have seen it when visiting a refinery in Turkey. It was printed at each facility, at each section of that refinery.

A.R.: By the way, I believe you have some brochures or booklets describing your companies and the facilities which were constructed there. I'd like to show them to our prime minister and some other VIPs.

M.D.: And then, I think they would like to have a conversation with you.

V.P.: We are terribly sorry. We gave all of them away before we came here.

What a mistake! No meeting with the prime minister took place.

Rule 1–38
Bring plenty of information advertising your company.

The advertising literature should be written in the English and Russian languages. The descriptions should meet some specific local requirements.

A.R.: We may want to have one large modern refinery. But since the sulfur content in our oil varies in different areas, maybe you will find that a better idea is to have three small modern refineries. As I have told you, in the eastern fields we have oil with up to 3.4% sulfur and westward it changes to 5.1% and more.

C.E.: Do you want us to tell you what you will need in this case?

M.D.: Yes, we do. We'd like you to prepare a feasibility study on this case.

V.P.: Will you pay for it?

M.D.: You will pay for it. Then you will have exclusive rights to accomplish your program here.

V.P.: Well, and if this program isn't approved by your government, we have lost our money, haven't we?

M.D.: It is your risk.

A.: I think your government has to show that its approach is serious. I'd suggest that we form a team to include your specialist paid by your government. Their duty will be to collect all the information needed for this feasibility study. And the American's group will do the rest.

A.R.: Will they compare different opportunities?

V.P.: We will submit the best choice.

This company specialized in constructing and operating small refineries; therefore, they knew exactly which version to choose.

A.R.: Will you compare the economic efficiency of each of the versions?

A.: Yes, they will.

C.E.: Unless this is a job for many years, I could solve this in a few equations if the amount of uncertain factors are limited. In a decade or so, these small refineries will be most profitable for sure. If we are talking about 20–30 years, probably a big one would be more efficient since a hydrodesulfurization is a very expensive facility. It is your choice.

A.: We will try to make the best choice together.

My initial point of view was that the same local team, which is working together with the American specialists on the feasibility study, will be responsible for final approval of the versions for the program (since there were no more refinery specialists in Tatarstan). Due to their personal involvement in the program, these people had to choose the version of small refineries since the problem didn't seem to have a single solution.

A.R.: Well, then I will give you all the characteristics of each sort of our oil . . . Do we have some kind of a protocol to be signed?

V.P.: Yes, we have.

M.D.: Albert, you have signed a few hundred protocols, haven't you. And after just a few bottles of vodka had been drunk. Did it work? I think the Heads of Agreement[6] should be signed now. And this time, vodka shouldn't mark the finish of a deal but the very start. You are with me, aren't you? . . . And also, gentlemen, let us mark this occasion by celebrating the birthday of our friend, Albert . . .

A.R. Well, before vodka, let us sign the Heads of Agreement. Do you have them prepared?

V.P.: No, we don't. We are ready to write it now and work all night, if it is necessary.

M.D.: Not now, not now . . .

What a mistake. The document was never signed.

Rule 1–39
Bring with you the documents that need to be signed.

These documents should be written in the English and Russian languages. Also, a good idea is to send these documents to your partners before your arrival. Even though your partners aren't use to this business procedure, they will be flattered by the fact that you treat them on an equal basis as your western partners. Try to start with a binding document; the time of protocols passed away.

(After some vodka was drunk):

V.P.: Yes, I understood. . . . Within the second week of May you'll get a detailed description of the refineries.

M.D.: I will show it to our specialists here. As you know, the city of Bougulma is the real center of our oil industries. So I will go there by myself. I have friends there, believe me. They need a week or so to write their opinion. They should be interested, of course, and if they are they will work fast.

Was this last statement a gentle hint of a bribe?

6. The Heads of Agreement is like a preliminary contract and is a binding document. Protocols are not binding documents.

M.D.: I'll take your description of the refineries when I am visiting Turkey next month. We have a joint venture in Ankara. I will discuss the financing of the refineries with my Turkish partners. Then the deal will go on.

> **Rule 1–40**
> **Tell your partners from the very beginning: I will never pay under the table, but I am willing to pay lavishly for each of your (or your friends) services.**

Add: everybody will know that you are paid; nobody will know how much.

V.P. I was in Istanbul a few weeks ago. The Blue Mosque is . . .

Never pretend that you did not catch a hint of a bribe. It is a matter of an obligatory response. Otherwise, your partners will disregard you.

C.E.: By the way, do you have a Mosque here in Kazan?
M.D.: Yes, we have. There was a time when a Moslem wasn't permitted to live in Kazan. This time has passed away forever.
C.E.: Are you visiting the Mosque?
M.D.: Time to time.

Be careful asking questions about their beliefs. All of your partners here (no exceptions) are former members of the Communist Party. For them these matters are unpleasant. Although most of them are reticent Moslems, it is an expression of their national feelings rather than of belief.

◆

The promised description of the refineries wasn't sent to Kazan in time. It was sent by fax directly to Istanbul, when Mr. Dautov visited Turkey. It was in English and neither Mr. Dautov nor his translator could understand it in order to explain the deal to the partners. When he brought the description back to Tatarstan and had it translated, it appeared that the described refineries are meant for . . . sweet Texas oil.

◆

"They don't respect me at all. I thought that Americans are accurate and reliable. Now, if they are like our people here, we should treat them in the same way. Look, I was sure that in America such companies wouldn't survive one day. How did they?"

(M. Dautov, Minister of Community Services of Tatarstan; conversation with author, Moscow, November 1993)

"These guys are just fools. If something happens they become wild, and it is impossible to solve any problems with these wild men. They never see their own mistakes but each one made by others. If I have bad relations with you but we have a deal—okay, let us make a deal. No way in there . . . "

(Vice President and CEO of an American oil corporation; conversation with author, New York, December 1993)

In August 1991, during the *coup d'etat* against M. Gorbachev, a delegation from Yakutia (an autonomous republic in east Siberia) visited Denver, Colorado. During the usual banquet, the prime minister made a toast "to the great success of the new government of the Soviet Union." The Americans kept quiet. One of the ministers[7] (not a Yakut but a Tatar) stood up, pushed his chair out loudly, and left for the lobby. There he made a speech in Russian.

"*Sekir bashka.*[8] But I am not wild enough to cut their heads off. Why, my Lord? They were our slaves for hundreds of years, and they are still slaves. But we aren't wild enough to make them our slaves and cut their throats any time we wish to. We are old, and our energy, also."

The Communist elite everywhere in the FSU didn't like and support M. Gorbachev's reforms. Some clever people of the same elite did. The minister said that his prime minister likes to remain a slave of Moscow rulers rather than become free.

---

7. Permission was not given to use this minister's name.
8. Turkish for, "Cut his head off."

"You have told me this story and I, also, remind myself of your words about the old energy of their bitumen. Do you want to know what came to my mind? Maybe our energy has become old. Man's power—nothing else—is the key there, in Eurasia, as it is everywhere. It seems that the people there have lost this key. As a foreigner, have you the impression that American energy has also become old?"

(Senior Process Engineer of a large American engineering company; conversation with author, Moscow, April 1993)

Chapter 5

Moslem Belt of the Former Soviet Union: Rules, Comments, Advice

For readers' convenience, the rules described in Part 1 are collected together in the beginning of this chapter. The rules are not given in the numerical order of the previous chapters but by type and order in which they would be used by a company establishing business relationships in the FSU. Advice and comments on everyday necessities of life are presented after the rules.

Be guided by these rules. You won't reach success automatically; by ignoring these rules, you will get into trouble automatically.

Rules

Introducing Your Company and Yourself

Be introduced to local partners by a Moscow VIP. Come to visit your local partners after you are invited by them.

The chain (of people) may be long until you reach the final person who will introduce you; it doesn't matter. Reference letter(s) may be useful; send them first to obtain in writing an invitation to visit the country. This is your local agent's duty also. If recommended from Moscow, search out the nationalities of both persons: the person introducing you and the person to whom you are recommended. It is better if the latter is of the local nationality. (Rule 1–37, Chapter 4)

Let rumors of your company's wealth and generosity run before you.

It is your local agent's duty. Supply them with the best advertisement you have. Wherever you are, never use community transport (buses, trolleys, etc.)—rent cars or vans; rent helicopters to see sites (deposits and oil field areas); dress in your best style. Some exaggeration is permissible. No greediness is pardonable. (Rule 1–14, Chapter 1)

Bring plenty of information advertising your company.

The advertising literature should be written in the English and Russian languages. The descriptions should meet some specific local requirements. (Rule 1–38, Chapter 4)

Always bring with you the documents that need to be signed.

These documents should be written in the English and Russian languages. Also, a good idea is to send these documents to your partners before your arrival. Even though your partners aren't used to this business procedure, they will be flattered by the fact that you treat them on an equal basis as your western partners. Try to start with a binding document; the time of protocols passed away. (Rule 1–39, Chapter 4)

Use, as often as possible, the expression: "We are here to help you." Believe it.

Discuss the social results of the mutual activity; i.e., construction of hospitals, schools, mosques; training of the local personnel; creating new jobs, etc. Show your personal attention to these activities. Some exaggeration is permissible. (Rule 1–23, Chapter 2)

Advantages of the Personal Relationship

Establish a personal relationship with local people while you have no problems, because it makes this relationship reliable; use a personal relationship when you face problems; it strengthens this relationship.

Within the Western culture one is a partner and then may become a friend. Within the Oriental culture one is a friend and then may become a partner. (Rule 1–1, Chapter 1)

Use personal relationships to search into the background of your intended interlocutor to determine the identity of his real (not formal) boss. Appeal to this boss when having problems in negotiations.

The Oriental hierarchy is far more complicated than the Western, where most often the formal and real boss is the same person. Be sure that in central Asia the reports sent to a formal and real boss are quite different—as their orders would be different. (Rule 1–2, Chapter 1)

Search the firmness of the position and the situation around your main partner.

If he isn't a newly appointed person, he may well need your friendship more than you need his. (Rule 1–24, Chapter 2)

Etiquette

Business is nothing more than the consequences of personal relationships and etiquette.

The business interests (and the mutual interest) are far secondary since these peoples mostly have no money interest in the deals for which they are responsible. (Rule 1–10, Chapter 1)

Your greeting should emphasize both the name and the position of the person with whom you shake hands.

Oriental people use this greeting among themselves also, but in very special cases. Make this case special. (Rule 1–9, Chapter 1)

Listen.

Sometimes listening is boring, but the success of the negotiations most often depends on your ability to listen. (Rule 1–19, Chapter 1)

Spend at least the first 10 minutes of a meeting in simple, non-business conversation. Wait for the host to begin the negotiations.

In order to understand whether the negotiations have or have not started yet, you have to know the host's background. (Rule 1–18, Chapter 1)

Show your willingness to be a pupil.

It is against your ego, of course. But in the Oriental cultures the relationship between the teacher and a pupil is most respectable, and your teacher becomes your defender everywhere, also. Your time to teach will come soon, immediately after you have established a business here. (Rule 1–6, Chapter 1)

Don't expect punctuality from the local VIPs.

An Oriental VIP must perform willfully. But you follow this French proverb: "La ponctualite est la politesse du Roi." (Translation: The punctuality is the politeness of a king.) Be a king. In time they will respect you. You have nothing to lose by following this proverb. (Rule 1–35, Chapter 4)

On Women

Never mention wife or daughter(s) while talking. [1]

Wives and daughters are esteemed in American culture; however, in the Oriental culture they are just a taboo—not to be mentioned. If you want to insult someone, simply break this taboo. Of course, not one local person thinks that a foreigner should know their laws and customs but, naturally, a special kind of behavior is expected from a guest. The question: what is your income (or wealth) cannot be accepted by an American, who would consider it indecent. Does it matter that he was asked by a foreigner? (Rule 1–11, Chapter 1)

You may discuss the beauty of local girls. Then be prepared that some of them are invited to a banquet especially for you. You should do the same when your partners are in your country.

No women with you at *Hoshars* (places where men meet). No exceptions.

No comments. (Rule 1–26, Chapter 2)

Their Personal Interests

Ask your partners which gifts they want you to bring them. Bring the gifts.

The first gift should be just a souvenir; the next time—the more expensive your gift is, the better. (Rule 1–16, Chapter 1)

Invite your partners to visit your country, and organize that as soon as possible.

In your country, you will be amazed at meeting such changed persons now willing to discuss business in the best way. (Rule 1–4, Chapter 1)

Let the local people pay your bills (restaurants, rents, hotels, phones, etc.) if they insist on doing it.

They pay rubles, you return their expense to them in currency (dollars). This is a big deal for them. (Rule 1–15, Chapter 1)

---

1. This rule is less important in Tatarstan.

Be aware that easy chatting can turn out to be the most important part of the negotiations, if the host would let you know his specific interest.

It will be expressed directly almost never. But since you have met his interest positively, you have relocated your deal into the field of etiquette; now it is his turn to be positive. He is obliged to say *Yes* regarding your wishes. He has no obligations to follow his *Yes* yet, but from now on his negative reactions are restricted. (Rule 1–20, Chapter 1)

There is a long distance to go between signing a contract, receiving money, and initiating activities on these projects. Be patient.

A contract means almost nothing in this part of the world since a system of laws and courts, which would make the agreements binding, doesn't exist. A contract expresses a balance of interests of each of the partners. While the state was strong, its representatives had to follow the state's interests. Since the state became weak elsewhere within the FSU, its representatives are looking for their own profit in each deal. (Rule 1–33, Chapter 3)

Discussing Politics

Never initiate discussion on political themes. Discuss politics as long as your host wishes to do so and even longer, if your host started it.

Political issues are extremely sensitive for any manager you meet in the FSU. And within central Asia it is still dangerous to discuss these issues with foreigners. However, if the host starts the discussion, go ahead, even if you are in hurry. Remember, for him it is more important than the current business. If he discusses politics with you, it means that he has already approved your deal. (Rule 1–5, Chapter 1)

Avoid participation in any discussion about international relationships.

If you join local opinions, they won't believe you; if you raise an objection to it, they will seek to kill you. (Rule 1–29, Chapter 3)

Discuss the political leaders of the country only after your host has mentioned them first. Discuss only the leader who was mentioned.

In the Oriental cultures the *Ruler* was, is, and will be sacred forever. One can hate *Him*, kill *Him*, and still worship *Him* as a divine creature. (Rule 1–7, Chapter 1)

Avoiding Troubles

When starting a business, try to keep the appropriate Moscow authorities at least neutral.[2]

It doesn't mean that you need to inform them in advance; it means that the source of their information must to be chosen by yourself. Your partners may decide to report to Moscow in an especially negative manner; prevent this from happening. Also, it is a good idea to be introduced to local people by a Moscow VIP. (Rule 1–36, Chapter 4)

At the negotiating table work as a team.

At the table, discuss any problems among you openly and loudly, and the translator may explain to the hosts the matter of your discussion. But no contradiction is acceptable in the statements of the different members of the delegation to your hosts. (Rule 1–3, Chapter 1)

Don't push your interlocutor when you face resistance.

If you do, later you may expect their resistance and obstruction, even if a direct order to meet your interests has been given by the government. (Rule 1–12, Chapter 1)

Try to set up a second meeting to discuss the problems which were not solved during the initial meeting.

If you succeed, then nothing is lost. But the second meeting is the last chance; you should know exactly how to make your conversationalists change their minds. (Rule 1–13, Chapter 1)

2. Less important in Uzbekistan.

If your partner is of a local nationality and his background includes education outside of the local region, you should be aware that he doesn't understand the deal you are negotiating, since he is likely to be ignorant in the technical and financial details of the negotiations.

This individual will never confess that he doesn't understand something; he would try to avoid any details which might be important to you. He wouldn't like you to insist on clarifying those details, and the negotiations will collapse. Try to have his deputy and some engineers at the table. (Rule 1–32, Chapter 3)

Avoid appointing a specialist of a local nationality to a position which requires knowledge and responsibility in your joint venture.

Especially avoid people who graduated from the Moscow or other Russian high schools since they were involved in the equal opportunity game. Those, nevertheless, might be useful for keeping warm relations with the local establishment. (Rule 1–31, Chapter 3)

Saving Money

Give as much information as possible about your activities in the country to representatives of the *different* clans. Try to make them compete to become your ally.

Since you made this kind of competition rational (i.e., where profit is involved and not ancient animosities), it should be much easier to manage the situation. This will also cut the possibilities for them to extort bribes from you, since they are watched by other clans' men. (Rule 1–8, Chapter 1)

Try to avoid paying money to any governmental subdivisions until you have signed a contract.

This is a way to be respected. (Rule 1–21, Chapter 1)

It is very easy to violate any agreement anywhere in the FSU.

This rule means that it is possible for foreign partners to play such games and to escape with minor troubles. However, don't misuse this possibility. (Rule 1–34, Chapter 3)

Your Local Agents

You should hire a local interpreter for the introductory part of the conversations with VIPs.

The local translators are not qualified to interpret technical details. But they would automatically improve your mistakes on local etiquette. (Rule 1–17, Chapter 1)

Your local agent(s) should have access to the Hoshars (or other men's parties) at different levels (including the elite's neighborhoods).

And it is better if this agent has enough power to invite you there. He will charge you $5,000 or even more. It is worth it to pay this bill. (Rule 1–25, Chapter 2)

Never fire your local agents.

Think before promising a local agent commission money. And, you should strictly carry out your promises. Keep your local agents neutral, out of the deals, if you don't need them anymore. Whatever you do, don't fire them because they will surely become your dangerous enemies. (Rule 1–27, Chapter 2)

Safe Behavior

Avoid using the regular local flights.

Rent a car with a driver. Rent a helicopter or other aircraft when you are in a hurry. It is cheap enough and, of course, much safer. If you pay currency, they will find fuel and other supplies. (Rule 1–22, Chapter 2)

While traveling to the republics outside Russia, never visit a physician who is a native of that republic.

The chances that a physician of a local nationality bought his diploma are huge. Seek a doctor who is Russian, Jewish, or Armenian; they have never participated in the equal opportunity game. (Rule 1–30, Chapter 3)

Never follow the girl you have met to her place. You may be falling into a trap.

(Rule 1–28, Chapter 2)

Bribery

Tell your partners from the very beginning: I will never pay under the table, but I am willing to pay lavishly for each of your (or your friends') services.

Add: everybody will know that you are paid; nobody will know how much. (Rule 1–40, Chapter 4)

Theoretically it is possible to avoid bribery, but then you enter into the treacherous world of Byzantine diplomacy (spreading rumors, using clan-on-clan control, etc.). Be sure you are skillful and lucky.

Never pretend that you did not catch a hint of a bribe. It is a matter of an obligatory response. Otherwise, your partners will disregard you.

Advice and Comments

Hotels

Within the described regions, hotels are poor and bad enough but rather expensive (up to $100/day/room and more). They are especially bad in Grozny and Kazan. The former Intourist's hotels in Tashkent, Bokhara, Samarkand, and Almaty are better, but forget about services which are common all across Europe and America (soap, shampoo, and toilet paper in the bathroom; air conditioning; laundry; CNN on TV; messages; beverages or meals in your room; etc.).

Prostitutes and thieves have almost free access to the hotels. Be careful! Don't leave your money, credit cards, or documents in your room, even in the daytime, and always carry the room key with you.

The described areas (except Tatarstan) are in a zone of seismic activity. The hotels have earthquake warning sirens. Don't ignore these sirens.

If your local agent proposes an apartment for your stay, don't expect it to be more comfortable or safer than a hotel.

The best place to stay is in a governmental *dacha*—a special hotel for the government's guests. Though it is more expensive than a hotel or apartment, a *dacha* is clean, safe, comfortable, and convenient. Also, it is most prestigious.

Meals

The Oriental cuisine is delicious. Enjoy it each time and anywhere you are invited by your hosts (a party at home or in a restaurant, at a picnic, at a *Hoshar*, etc.). Don't touch any food in the streets, at a bazaar (market), in a store, and so on.

There are three levels of public catering in the areas of the FSU. The first level is the dining place (cafeteria, *Stolovaya* in Russian). Never enter there. The second level is the cafe (the same word in Russian). You may have a glass of water or juice there. Watch and make sure that the glass has been washed and is clean. Most often it has not. Don't ask for tea or coffee there. The third level is restaurants. They certainly will not serve you if you enter only for a cup of tea or coffee. This is a good place for having your breakfast. Don't order any meat, sausages, or fish.

Have your lunch in places where you are invited by your hosts. Have your dinner in your hotel.

If your agent proposes a special cook for preparing your meals in the city and for your trips to sites, accept this idea since this is the best solution of this problem. Your food will be absolutely delicious and fresh.

Communications

Phones are available almost everywhere in the cities and at the sites you are visiting. The phone communications are reliable: you can call the United States and Europe and receive calls from the United States and Europe. Sometimes you may need patience while dialing the number you need.

Communications through fax machines aren't as easy since most offices do not have this equipment.

Federal Express services are available only in Moscow; United Parcel Service works only for the capitals of Uzbekistan, Kazakhstan, and Turkmenistan. The local post office service is absolutely unreliable.

Entertainment

Once more, the Oriental people are extremely hospitable, and your hosts will do their best to entertain you in the evenings and through the weekends by inviting you to their homes, to the restaurants, and to picnics, etc. Don't refuse: it is their real pleasure and not an obligation. And it will become a pleasure for you to watch them in informal situations.

Besides, in the fall and winter, you may visit their ballet and opera theaters which are still in good condition. You may especially enjoy their national ballets and operas with very exotic music and dances. You can be assured that they have first class dancers and musicians in their theaters.

You may visit their rich historical museums and art galleries, and you'll never be sorry for the time spent. The capital of each republic has these facilities and after visiting them you'll know something new about the people you are dealing with.

Shopping

Their bazaars are rich and wonderful. You will never forget the taste of local fruits and vegetables. These need to be washed thoroughly before eating.

You may buy some souvenirs at the bazaars, also.

Have the local currency on hand; don't change your money anywhere but in your hotel or with your hosts personally.

Their stores and shops are poor. If you want to buy a specialty item of local production (silk, carpets, etc.) make a request to your hosts. They will bring you to special peoples and places they know, and you'll have the best prices.

Don't buy in black markets without your hosts. It is dangerous: first, because the items you buy will be of bad quality; second, you might be robbed; third, the police will appear out of nowhere and you'll have to bribe them.

Special Notes

Although the climate is extremely hot, neither you nor the women of your delegation should ever wear shorts anywhere, including your hotel.

The women of your delegation should only wear formal dresses. These dresses have to reach their knees (not shorter) and have short sleeves. Sleeveless blouses and dresses are not acceptable. Jeans are permissible at picnics only.

The women of your delegation should never speak or laugh loudly.

When entering a house (apartment), always ask whether or not you have to take off your shoes. The answer will be: not at all, but you *will* take them off.

Be especially attentive to the elders. Keep silent while an elder is talking and as long as he talks. (The longer he speaks, the more respect he shows to his guests.)

Part 2

This Is Russia

Introduction

In 1993, between Russia and the United States, the commodity turnover was more than $2.5 billion. Russia exported to the United States precious metals and jewelry, non-ferrous metals, chemicals, foodstuffs, oil, and oil products. Russia imported mainly foodstuffs (wheat, corn, barley, butter, beans, etc.), machinery, and equipment. The Russian-American balance of trade has a long tradition of a deficit: since 1952, the annual costs of imports have been as much as four times more than the costs of exports.

In 1993, both sides encouraged commercial contacts, an increase in the turnover of commodities and services, the trade of equipment and technologies, and the creation of small- and medium-sized ventures. Two important agreements were signed: on encouragement and reciprocal defense of investments[1] and on non-double taxation.

In Russia, U.S. companies are the leaders (among Western partners) on creation of joint ventures. (More than 1,400 joint ventures have been created, and more than 300 American companies have representatives in Moscow.) But in 1993 the total American investments in Russian industry was less than $400 million. (Total Western investment was $2.9 billion.)

The Western investments were dispersed as follows:

- Machinery construction and metal-working industry: $678 million (23.4%)
- Oil and gas industry: $475 million (16.4%)
- Public catering and food industry: $443 million (15.3%)
- Building construction: $156 million (5.4%)
- Woodworking and pulp and paper industry: $132 million (4.6%)
- Building materials industry: $111 million (3.8%)
- Others (trade, medicine, tourist industry, etc.): $905 million (31.2%)

1. This agreement was not ratified by the Russian Parliamant in 1993.

The Western investments were dispersed among Russian regions as follows:

- Moscow and Moscow region: $794 million (27.4%)
- Krasnoyarsk region (Siberia): $414 million (14.3%)
- Omsk region (Siberia): $238 million (8.2%)
- Arkhangelsk region (northwest of European Russia): $231 million (8%)
- Jewish autonomous region (Russian Far East): $173 million (6%)
- Belgorod region (Central Russia): $119 million (4.1%)
- Mary Republic (Central Russia): $145 million (5%)
- Komi Republic (Northern European Russia): $106 million (3.7 %)
- Others: $680 million (23.3 %)

In the oil industry, almost 40 joint ventures were created with American partners. Some of them were developing almost 50 oil and gas fields in different regions of the country, and they had produced more than 8.9 million tons of oil in 1993. Eight joint ventures provided various services such as formation fracturing, acid frac, sand frac, workover, etc. In 1993, these joint ventures produced 7.4 million tons of oil.

In 1993, Russia exported (outside the FSU) 79.7 million tons of oil in total, and 7.69 million tons of this oil was exported by the joint ventures. In 1994, the total production of joint ventures is expected to increase up to 25.5 million tons and their export—up to 15.9 million tons. The shares of the American partners in these joint ventures varied from 31% to 50%.

So far, most Western and American deals have been small, but the time for bigger projects is coming; eventually, the Texaco project in Komi will be signed, the Royal Dutch Shell and Amoco projects in western Siberia and the Sakhalin projects are to be signed after an often tortuous three to four years of negotiations.

> "This can be a trying place. The investment climate now isn't even as good as it used to be. But it's a matter of having patience and working constructively."
>
> (A. Morland, Vice-President, Amoco, interview with a U.S. newspaper reporter, March 20, 1994).

The events that occurred in Russia—and especially the way they were covered by the American news media—confirmed for me that foreigners need to know much more about the country they are willing to deal with and from which they expect a great profit.

To this end, I believe that the stories about the FSU manner of doing business would serve as guidance; therefore, they are included also. In the chapters below, some seemingly less important features are described in more detail than some seemingly more important features. But there is a good reason for this presentation. Foreigners are more familiar with the unchangeable features—such as geological structure, system of management, political directions, or the ploys of satisfying personal interest, etc. The changeable features of modern Russia—either attempts of democratization, or attempts to attract foreign investments, etc.—might contain some details which would probably survive, and these will become the real fulcrum for some kind of long-term business relations with Russians. It is the details which will survive that I tried to emphasize in the following chapters.

In the beginning of 1994, the Gallup Service's poll made for the Control Risks Group (England) showed Russia as the most difficult country for creating and conducting business (African and some South American countries were next in this poll).

I'd suggest that these are the reasons which form this opinion:

1. Political instability and a potential for civil wars
2. Weakness of President Yeltsin's government
3. The increase in criminality and organized crime
4. Terrorism
5. Commercial and financial swindle
6. Corruption of authorities at every level of government (central and local)
7. Uncertainty of the hierarchy of industrial management
8. Unfranchised status of foreign investments and investors

No ready recipes exist for dealing with these obstacles. Examples, rules, comments, and advice are given below for supporting your imagination and creativity. Imagination and creativity are especially important for oil-related business since it demands large and long-term investments.

I wouldn't say that the Russian oil industry is in the process of self-destructing, but it is disorganized and damaged by incompetent and inconsistent reforming. The chain of the oil industry consists of these main links: seismic survey, exploration, development, refining, transportation, and shipment of oil to the world market. In the FSU, these links are broken, and each link faces its own problems. These are described in the appropriate chapters along with some potential areas of investments. Also included are several simple calculations which show the scale and proportions of investments and the expected profit.

Although some of the results of the conclusions seem to be negative, it doesn't mean that they need be negative forever. Hopefully the economic changes in Russia will become related to the world market.

It is true that it is easier to destroy than to create. It is true regarding the iron curtain also. It was difficult to destroy the iron curtain, but it would be almost impossible to create it once more.

> "If Russia is to become a serious partner in building a new world order, it must be ready for the disciplines of stability as well as for its benefits."
>
> (H. Kissinger, former U.S. Secretary of State; March 14, 1994)

If the new world order is to bring benefits, Russia must become a stable partner. And this new order has rather a short period of time to find the proper keys to open Russia's still closed storerooms. These storerooms are rich enough to benefit Russia as well as to support the stability of this new world order.

Russia must see something different, not exactly what President Bush preferred to see in the new order; also, the new stability inside Russia will not satisfy President Clinton's expectation.

The opinion that Russia is stable only if closed is supported by its history from 1917 to 1987. Before 1917, however, the country was stable and open for 170 years, and during these times foreigners and foreign investments were welcome. Hopefully, this open historical model will be realized in the near future. Then, the risk will be lower and the profits will be more evenly spread.

But by then, it will become much more difficult to penetrate into Russia's enormous market.

Chapter 6

Moscow's Volcanoes

> "I like your socialistic ideas. In order to realize them we should find a nation which nobody cares about."
>
> (O. von Bismarck, Chancellor of the German Empire; excerpt from a letter to K. Marx, 1862).

Every dictionary and interpreter would translate the word *money* into Russian as *dengy*. All of them are wrong since no analogies exist in modern Russian for this term *money*.

For a Westerner, this term *money* is the measure of a few dozen things like cash, credit, credit card, credit line, bank account, investments, profit, assets, real estate, income, wages, royalties, etc. In post-Gorbachev Russia, the word *dengy* means immediate cash — nothing more.

Rule 2–1
Use proper financial terms when discussing business. Avoid using the word money. Use the word cash when you have this in your mind.

This is the way to avoid a lot of confusion and misunderstanding.

Just after they took power in 1917, the Communists — following their great ideas — tried to eradicate forever the *evil* money. It wasn't an easy task, but they didn't lack the resolve to accomplish it. Even in the period of the New Economic Politics (1922), when private entrepreneurship had been permitted, V. Lenin advised that the banknotes be printed on poor quality paper (so as not to survive for a long period of time).

A new feature had been invented in 1927: a not-to-be-cashed ruble.[1] This idea was very simple and worked for more than 70 years: the price on each item produced (or on shipment by railway or on different services, etc.) was appointed by the government. These and only these prices were in use between the state-owned enterprises (after 1927, no other enterprises existed). Thus, the incomes of each deal (or losses, profit, real estate costs, taxes, etc.) was counted in specific figures, which were registered in the local state bank (no other banks existed). Furthermore, none of the state banks would cash this money, ever, since it was—from the very beginning—nothing more than a number of large figures. It was called *dengy.*

The Communists tried to do the same to the people (workers, peasants, etc.). Workers would obtain a document which confirmed that they spent eight hours within a factory (a lab, a field, etc.) each day during a week and then—by this document—a store (state owned, of course) would give them everthing they needed for living. But this idea didn't work, evidently.

Then a special, strictly limited fund was appointed to each enterprise; only the figures appearing in this fund were available for being cashed. Thus, money appeared once more in society, but you could use this money only to buy the absolute necessities of life. For buying anything extra (foreign goods, a car, an apartment, a house, etc.), this money wasn't enough. Special permission was required. And additionally, the price for the same item (a car, a house, etc.) was different and much higher when it was paid in cash (which was used only for private purchases).

In the late 1980s, M. Gorbachev tried to make these figures in the banks available for cash. Billions of rubles had to be printed, and inflation had begun. He lost his presidency; the FSU ceased to exist; Russia and the Commonwelth of Independent States (CIS), together with the other former socialistic countries, are going through the worst inflation ever seen in this part of the world.

Why did M. Gorbachev let it happen? He believed that the figures in the banks were real money. But it was *dengy* (not-to-be-cashed).

B. Yeltsin's government could save the ruble only by supporting it with the dollar. So the ruble became immediate cash.

1. This was the Russian national currency for ages. The name comes from the Russian for "a stump (a piece) of silver."

> "No such crazy man exists here who can predict the fortune of the ruble. But this country is so rich with mineral resources, for example, that its own currency will survive automatically. What does it mean — automatically? It means that the market needs it, and the market will support it."
>
> (G. Yavlinsky, Economic Adviser, Member of the Soviet and both Russian Parliaments,[2] in an interview, Moscow, December 1993)

Let us take a quick look at these mineral resources which support Russian currency. The figures below were given by V. Orlov, Chairman of the State Geological Committee during a press conference in the beginning of 1994.

The total cost of raw materials produced in Russia in 1993 was $170–175 billion.

The total cost of mineral resources discovered in Russia is worth $28,600 billion; and the estimated resources — $140,200 billions.

The country's provisions of different discovered mineral resources and oil was given for the coming years:

- Gold (in ore): 30 years (South African Republic—for 37 years)
- Gold (in place) 12 years (South African Republic—for 75 years)
- Phosphate: 52 years (the whole world—280 years)
- Antimony: 14 years (the whole world—75 years)
- Coal: 180 years (the whole world—400 years)
- Oil: 30 years (the whole world— 50 years)

At another press conference, the minister of foreign economic affairs said that in 1993, Russia exported goods worth $42 billion; almost 50% of these goods were receipts from oil and gas sales. The same year, Russia imported goods worth $24–25 billion which is a decrease of 52% compared to 1992.

◆

In the last five years, some people became very rich due to trade, especially those who seized greedily upon the oil trading. This had been the state's

2. The Soviet Parliament ceased to exist together with the FSU. The first Russian Parliament was dissolved by President Yeltsin in October 1993. The second Russian Parliament was elected in December 1993.

monopoly for years, and the oil money which supported Brezhnev's government was spent in a particularly irrational manner (to help the alleged allies in Angola, Somalia, or Libya, or in the USA, etc.) and wasteful way. These new owners of the oil money don't act any better. But, at least they have saved some of the money in Western banks and are trying to understand what to do with it now.

◆

> "I can tell you this: if only half of the money owned by some Russian citizens would come back home and be re-invested here, we wouldn't need any foreign investments at all."
>
> (G. Yavlinsky, Economic Adviser, Member of the Soviet and both Russian Parliaments; in an interview, Moscow, December 1993)

The reasons why these Russians don't reinvest their money in Russia's private (or state) enterprises are the same as why the Westerners stay away: no rules of this game exist, and nobody is strong enough to introduce and defend those rules.

In March 1994 some special rules for Russian exporters were introduced to prevent these exporters from hiding their money in foreign banks. Now these exporters must describe all the details of the trade in a special statement to their Russian bank. This Russian bank must submit the description to customs and control the return of earned money from Western banks. Customs has to be informed when the entire deal is over. This double control didn't improve the situation very much, since Russian exporters, describing the deal, were artificially decreasing the prices of their production.

Rule 2–2

Contact state (quasi-state) enterprises since the logistics and financial rules regarding their relations with foreign investors are still much more clear and reliable.

Quasi-state means all kinds of joint stock companies, associations, etc., which were transformed from the state's enterprises by changing the name—nothing more important has been changed in their status; they remain state-owned.

Rule 2–3
Contact those private enterprises which have been formed by the heads of state-owned enterprises for better opportunities for financial maneuvers and/or for personal business reasons.

Almost each state-owned enterprise has such "private" partners. Obviously, the friends, relatives, or people with proper influence manage these kinds of enterprises.

Rule 2–4
Stay away from those newly established private enterprises which aren't specialized yet and are looking for a niche for their activities.

It is almost impossible to determine what kind of "dirty" money they own if they do own any. Their contacts in the government might be based on bribery; plenty of agents representing gangs of racketeers serve therein.

An American company was looking for a buyer of two small refineries (almost 1,000 ton per day capacity) in Russia since these refineries proved to be economically inefficient in the United States. In Moscow, these Americans were introduced to the owners of a newly established private company which was looking for some business. Mr. V. Belov, president, was a former helicopter pilot in a special subdivision for carrying members of the government. Mr. E. Protcenko, vice-president, was a former chief of the Moscow regional chess club.

After a short conversation, the two gentlemen decided that they were interested in the refinery deal. They appeared in a week or so with the wonderful idea that a refinery should be installed in Moscow's International Airport (Sheremetyevo) to supply foreign airlines with jet fuel.

That evening, Mr. Belov was in the lobby of the hotel *Slavyanskaya* in Moscow. Being a little drunk, he signed a Purchase Agreement (in English) which obliged his company to pay the first $5 million within two weeks (otherwise the deal is void). The Americans were hopeful because he showed a few valid and proper documents from a Cyprus bank.

Then Mr. Belov disappeared for a few months. The Americans were told that he was looking for money (credits) in Cyprus, where his company had an offshore branch.

Meanwhile, one of his employees, Mr. Alexandrov, left for a private bank (Victor). The owner of this bank had spent a few years in an FSU prison for private enterpreneurship which was strictly forbidden during the entire Communist era. The position of this bank Victor was pretty confused, and they badly needed a hard currency credit line with a foreign bank. Since Mr. Alexandrov spent a few years in New York with his parents (UN employees in the late 1970s), he—as the bank's owner thought—could help. Mr. Alexandrov invited the Americans to negotiate the refinery purchase—having in mind, however, that they could help to obtain a credit line for this purchase and some additional amount of money for the bank.

As a mortgage for this credit, bank Victor proposed a few of their new aircraft liners, insured with Lloyds of London. This deal could be financially restructured in a Western country, but this was Russia and the Americans reasonably decided not to enter the deal, even though they badly needed to sell those refineries as soon as possible. (Their money was frozen in those refineries and the maintenance had to be paid regularly, etc.)

Meanwhile, at the end of summer 1993, Mr. Belov and Mr. Protcenko appeared once more with a new location for the refinery in an area at a distance of 150 kilometers north of Moscow. They proved to be supported by local authorities; by Moscow's regional governer; by Mr. A. Karpov—famous former world chess champion and a clever businessman; by the chairman of the State Investment Committee (who was close to President Yeltsin); and by Mr. Burbulis, President Yeltsin's *grey cardinal.*[3] In a letter, the State Investment Commitee promised its support for Mr. Belov's company to get a credit guaranteed by the government of Russia.

A new Purchase Agreement was signed obliging Mr. Belov's company to pay 20% of the cost of the refinery within one month. This money was never paid, and the refinery was never sold to this private company.

---

3. A *gray cardinal* is a powerful man who makes presidents, prime ministers, ministers, etc. He isn't involved directly in events, but he is ruling the country from a political shadow. This expression comes from France where Catholic cardinals were very powerful but didn't appear openly at the political scene.

"I would tell you this. As you know, I was a member of the Central Commitee of the Young Communist League. This is how matters stand: while my former colleagues are alive, I am not investing one ruble in a private enterprise."

(A.E. Karpov, former World Chess Champion; conversation with author, Moscow, June 1993)

◆

It is true that these refineries mentioned previously were never put into production in the United States, but the facilities had been manufactured in the mid-1980s. The price appointed for them to be sold in Russia was twice as much as in the United States. The owners tried to deal with private companies in Russia relying upon the fact that those Russians weren't experienced in obtaining information about world market prices. This guess turned out to be right, but the deal was postponed due to different unrelated circumstances. (The first Russian Parliament refused to allot money to President Yeltsin's State Investment Commitee; then this Parliament was dissolved; then this commitee was abolished; etc.)

But the future crash of this deal wasn't accidental.

This American company had hired a few Russians (some of them living in the United States) to work with the company. Those who lived in Moscow were irregularly paid small salaries (on the American scale), but at least they were paid in cash. Checks sent to Russians in America were irregular also. However, these checks bounced at American banks. And, it took time for the Russian workers to recognize this ploy.

◆

"Even those who had a written agreement with them suffered. I didn't have an agreement, so I lost this money forever. Each time they would tell me: it was a mistake, Tony. Well, I tell this story to everybody I meet traveling across the USSR with my new employer. And I have repeated the name of this company more than 10 times. I'd like each dog in Russia to know this name."

(Antony Reznichek, interpreter; conversation with author, New York, November 1993)

Rule 2–5

When working with Russians, use the same moral and ethical standards you use with your fellow Americans.

This isn't an issue of spiritual values or reputation only. In the Western world, the law helps defend your interests, and, probably you could work with somebody even though you can't trust him. Russians don't have this luxury. But they have the possibility to choose partners since competition is rapidly rising.

Russia isn't closed to dishonest people, entities, or intentions at all, but, on average, the chances for them to succeed aren't much higher than in America.

> "For Russia's sake, these bad laws there become balanced by bad observance."
>
> (A.I. Herzen, famous Russian writer, 1853)

No certain laws (as a system) as to foreign investments, rights, private property, and legal redress exist in Russia or other CIS countries. A few laws passed by the Soviet and first Russian Parliaments aren't bad, but they still are "balanced by bad observance." These laws, president's decrees, governmental regulations, local authorities' edicts—all of these usually contradict one another and very often just miss the common sense. And it is always unclear who has jurisdiction or who will claim it—local authorities or Moscow's bureaucrats.

One of these two strategies might be chosen in these circumstances: the first (a passive one), everthing which isn't specifically permitted is forbidden; the second (an aggressive one), my Russian partner and I are creating and following the rules of our business. The worst strategy would be to stay between these two.

> "In Moscow, they are very anxious if some money passes by their pockets. We have invented several ploys to pull the wool over their eyes. Follow me and don't worry about them."
>
> (V. Shnurov, President, KEDR; conversation with author, Tyumen, November 1993)

Rule 2–6

When doing business, follow the laws and regulations which have been chosen to be observed by your Russian partner.

This isn't a choice between some "good" or "bad" laws (or decrees); this is a choice of people who control the observance. Your partner knows these people better.

◆

The oil industry of the FSU was organized (on the national scale) by a rational and judicious scheme, which was functional but doubly controlled by the central government and local Party's committees.

Within each of the petroleum basins, there were as many local operating companies as there were administrative divisions. (Each of the oil companies were supposed to be under the control of the local Party's committee.) These production companies passed their oil to the Basin's Pipelines Management and were paid by once-and-forever appointed prices (in *not-to-be-cashed rubles*). The Pipelines Management reported to the Moscow-located Central Piplines Administration; the local operating companies reported to the Production Administration of the same Ministry of Oil Industry. Another administration was responsible for all the supplies (drilling rigs, casing, pipes, etc.). If the money received for the oil wasn't enough to repay the supplies, the Financial Administration of the same ministry should *find* this money, i.e., discuss the problem with appropriate bureaucrats in the government and get a confirmation in writing that the money will appear (where from?). Usually the money (not-to-be-cashed rubles) appeared easily if the company produced the planned (in Moscow) amount of oil. If the company was behind this schedule, the money would still appear, but a newly appointed general director would be in command.

The local operating companies were responsible also for exploration drilling, reserves growth, discovery additions, extension additions, secondary recovery, etc. They had to report to Moscow's Geological Administration on these responsibilities.

Local operating geophysical companies were responsible for well logging and seismic survey. These companies reported to and were paid by Moscow's Geophysical Administration of the same ministry. This order was changed in the late 1980s, when the geophysical companies were paid for their services by local oil companies. Of course, the money came from Moscow, and, of course, it was still not-to-be-cashed rubles.

The same scheme was in use regarding the gas industry; additionally, the Ministry of Gas Industry was responsible for offshore exploration.

Within the huge Siberian regions and northern portion of European Russia, exploration was conducted by the Ministry of Geology, also. But this ministry had no rights to produce oil or gas.

It was paid (by not-to-be-cashed rubles) for discovered fields (the sum was dependent on the reserves discovered) by one of the two ministries previously mentioned; if no oil or gas was discovered, it was paid on a different scale.

The larger the figures of not-to-be-cashed money, the more supplies of different kinds (apartments, cars, salary funds, etc.) could be obtained. So this money was worth being earned. Therefore, all estimations of the initial oil or gas reserves in a newly discovered field were exaggerated. Then, after this field was put into production, the figures of reserves would be underestimated in order to reduce the annual production schedule (planned in Moscow). Then this figure of reserves had to be exaggerated once more in order to obtain money (i.e., supplies, salary funds, etc.) for secondary recovery. There were some other personal reasons for these games with the figures of reserves (decorations, promotions to higher positions in the hierarchy, etc.) as was previously mentioned.

◆

Rule 2–7

If you are interested in a field, check the reserves figures; check the initial data and methods used to estimate the reserves. Check the number of wells drilled; check the number of productive wells; test the production rate.

Usually the shown materials are wrong and nobody remembers when and why the data were changed. Since the drillers' salary depended on the weekly footage, even the depths report might be wrong.

> "Well, we really had an annual production rate of up to 650 million tons. As I told them in Houston, it had nothing to do with the system. If you have plenty of oil just under the soil, or if you have no oil at all, the system doesn't matter. The system does matter when you have some oil somewhere deep. This is where we are now, but a more appropriate system is being developed."
>
> (G.A. Gabrielyants, Minister of Geology of the FSU; conversation with author, Houston, January 1991)

◆

The minister of geology had chosen the simplest way to explain the changes in the oil industry. As a true Moscow bureaucrat, he was sure that the local people have nothing to do with oil money but work for their salaries (large salaries on the FSU scale). Mr. Yeltsin's staff had a closer look into the events which were about to happen. In 1989 and 1990, while campaigning in western

Siberia, Tatarstan, and other oil producing districts, Mr. Yeltsin promised the local people (the local operating companies) that under his leadership, they would obtain full ownership of their oil fields and oil. Of course he won the elections; he partially kept his promises, and he received plenty of money from the oilmen (the oil industry was among the first which was permitted by President Gorbachev to cash those not-to-be-cashed rubles); he could buy some of the officers KGB, police and Army, and, finally, he defeated President Gorbachev by destroying the FSU in 1991.

Becoming the president of the Russian Socialistic Federation, Mr. Yeltsin already had a few allies among Moscow's old oil aristocracy. Then in December 1991, Russia left the FSU which, obviously, ceased to exist. The Ministry of the Oil Industry was among the first which was abolished in the beginning of 1992. Moscow's old oil bureaucracy had lost its power. Those who could forecast or guess the events and had joined Mr. Yeltsin made millions of dollars by trading in crude oil. These people became the new oil establishment and they are smart enough not to be concentrated only within or around the Ministry of Fuel and Energy, which took over the oil business after abolishment of the Oil Ministry.

◆

> "Nowadays Moscow makes its living trading crude oil and raw materials. It is time to stop this shame."
>
> (G. Ziuganov, leader of the Russian Communists Party; TV interview, Moscow, November 1993)
>
> *Was it a shame for the FSU when doing the same?*

Naturally, this newly formed establishment tries to gain control over the local operating companies and Russian Prime Minister Chernomyrdyn (former Minister of Gas Industry of the FSU) heads this party. Naturally, the old bureaucracy dreams of regaining control over the oil industry. (Mr. Chernomyrdyn keeps them as allies, also.) And the local operating companies have no wish to report to Moscow again. (President Yeltsin supports them in order to reduce his prime minister's influence and to keep the oil money for himself.)

Naturally, none of these three groups—the old oil bureaucracy, the new oil establishment, and the local companies—are ready to share their money and influence with others, and especially with foreigners. (This is the only thing they have in common.) But since each of them have different interests, they are in need of the support of foreign investors.

Rule 2–8
Although the representatives of each of these three groups may become your partners, it is preferable to stay with the local operating oil companies.

Being immediate producers, these local companies badly need all kinds of equipment and machinery, and they would try to avoid any middleman—Russian or American. Their needs have to be understood (if not respected) even in Moscow.

In 1991, an American company signed a preliminary agreement on developing a large oil-gas-condensate field in the Astrakhan region (northward from the Caspian Sea). The fluids contain more then 6.5% sulfur, and this had become a real disaster there. Soon, a new law passed the Russian Parliament requiring a tender for each mineral property. The owner of this company happened to corner Mr. Gaidar (former Russian acting prime minister) at the White House in Washington, D.C. and asked:

"I happened to sign this agreement before this clever law was passed. What is the status of this agreement now?"

"This law, like any other, isn't retroactive,"—was the owner's reply.

Later in 1992, this American visited Moscow and had a long and difficult conversation with the deputy minister of fuel and energy. He tried to explain Mr. Gaidar's point of view. So the deputy minister picked up the phone, asked to be connected with Mr. Gaidar, told the acting prime minister the story, and pushed the speaker button.

"Do whatever you want," the American heard Mr. Gaidar say.

"Stop," he said with anger. "This is not what I was told in Washington, Mr. Gaidar."

But the phone was already disconnected.

In the beginning of 1993, this American visited the city of Astrakhan on the same day that Mr. Chernomyrdyn arrived. The two gentlemen had a short meeting, and the American was told that the Russian government had some other plans regarding the discussed field.

In the beginning of 1994, this area was still a disaster, but no other foreign company would work there because the local operating company tried to stay with this American partner only.

Early in 1992, Texaco started negotiations with the local authorities of the Komi Autonomous Republic (north of European Russia) on the exploration and development of a large promising region within this republic. In April 1994, after almost two years of agonizing negotiations, the deal was approved by the Russian government, which was very suspicious of Komi's moves toward economical independency.

Four large American companies (Texaco, Amoco, Exxon, and Norsk Hydro) formed a new company, Timan-Pechora. They plan to spend almost $100 million for exploration of one of the largest oil fields discovered in this region—Roman Trebs, with reserves as much as 350 million tons of oil (2 billion barrels).

Rule 2–9
Avoid getting involved in political fights, but stay ready, together with your Russian partners and allies, to defend your lawful interests.

In modern Russia, at least an attempt has become possible.

> "I had a feeling that you wanted investments without investors there in Moscow."
>
> (Director of an American petroleum information company; conversation with author, Houston, February 1991)

> "I had dinner with Mr. Shumeyko (deputy prime minister of Russia)[4] last night. He was pessimistic and said that Mr. Chernomyrdyn wouldn't try to make a paradise for foreign investments in Russia. He supports this idea himself. But both of them understand that it is a hell now and that this is what must be improved."
>
> (Vice-President of a large American investments bank; conversation with author, New York, July 1993)

4. Mr. Shumeyko became the speaker for the Chamber of Russian Parliament in January 1994.

"Now, do you think that Yeltsin is really in charge there? We believe he is the only one in the Russian government who needs foreign investments in Russia. If so, we have less than two years for the money to turn over."

(Vice-President and CEO of an American oil company; conversation with author, New York, February 1994)

◆

In the beginning of 1988, Mr. Yeltsin lost his positions both in the Political Bureau of the Central Committee and in the Moscow Committee of the Communist Party. After General Secretary Gorbachev and his retinue heard the last speech of Yeltsin as a Moscow leader, and the required words that he " . . . was, is, and will be forever devoted to his dear Communist Party," Mr. Gorbachev left quietly, in vain.

◆

It isn't clear why General Secretary Gorbachev made this fatal (for himself) move.

At the end of 1988, Mr. Yeltsin was campaigning to become a member of the Russian Parliament.

One day Mr. Yeltsin had a 6:00 P.M. meeting with the workers and engineers of a large military plant in the northwest district of Moscow. However, the Party's District Committee advised the general director of the plant to let everybody leave at 4 :00 P.M. Of course nobody waited for Mr. Yeltsin, and the hall was empty. A rather small company where I previously worked was located next door to this military plant. So Mr. Yeltsin's aide happened to enter our door and asked whether or not we wanted to see Mr. Yeltsin. We wanted to.

Mr. Yeltsin's speech was very short (he was talking about a democratic society), but he spent four hours answering questions. These were my written questions handed to the aide (my handwriting is bad, and I made a mistake—I didn't print the questions):

Is it possible to establish democracy in a society without private property?

Is democracy possible in a country where one party is in power forever?

Is democracy possible in a country without rooted traditions of this kind?

Do you think that the the slower the changes in the society the more reliable they are?

When he picked up my list of questions from a pile on the table he could hardly read the first question aloud, and then he said, "Sorry, I cannot read this handwriting."

It was almost midnight when the aide ask me to follow Mr. Yeltsin's car and then—after escorting *the Boss*—to pick him up and bring him home (somebody had advised him that we were neighbors). So I started my car and was waiting for Mr. Yeltsin's aide when to my amazement Mr. Yeltsin got into my car. Soon we left for Mr. Yeltsin's house.[5] This was the short conversation in my car:

"I understand that you are very tired, Boris Nikolaevich."

"Don't use this word (tired). I don't like it. Better, you tell me your impressions. Maybe I'd like them more."

"Here. First, you liked this meeting and the people appreciated it. Second, you wished not or could not hide your hatred of Gorbachev. Third, I am terribly sorry for my handwriting; it was my fault, really, but I am curious what you had in mind when using this word: democracy."

"So you think that it has to be my fault." (introducing democracy)

"May I insist?"

"No."

Does the president of Russia think that he must find proper answers to the questions above? Or, this very tired man in my car was correct in thinking that it is rather the citizen's duty to find out the answers?

On March 7, 1994, President Yeltsin appeared on TV with a speech to Russian women. This was the first time he didn't use this word democracy at all. Instead he used and emphasized the meaning of "the state" for Russians. The state was, is, and will be the central idea of Russian history and life. At least, Russians have forced their president to make this choice. ***(It is better to say that he was shown that he has no option.)***

◆

A few years later I called Mr. Yeltsin's aide and asked him whether or not it was worth bothering the president with a specific deal. An American company had sold some drilling equipment to a Russian local operating entity,

5. Later, I thought this entire incident was a specific ploy to avoid the KGB's eyes since a few accidents had happened lately with Mr. Yeltsin.

and this entity had transferred a certain sum of money to a Russian bank which was supposed to transfer this money to the Americans. The bank didn't because of . . . lost documents (so it was explained the first time), and after a month or so this money still wasn't transferred because of . . . a lack of money (so it was explained the second time).

The Americans badly needed the turnover of this money for another investment.

◆

You can be sure that the money was turned over (once or several times)—but for somebody else.

The president's aide advised me by phone: "Don't bother the high political authorities with your business problems. Look for a proper key."

Rule 2–10
Don't appeal to the high political authorities with your business problems. Look for a proper key.

In modern Russia, it is almost impossible to tell who is who in different businesses (especially in the financial mafia)[6] *even for the highest political leaders. So the latter would be very careful in entering all kinds of business squabbles. The "proper key" means proper people and their interests.*

Rule 2–11
Don't expect a fast money turnover.

Even some trade companies got stuck on this issue, which became a real national disaster. It is difficult to improve this practice since several very powerful groups make their profit by the turning over of money which doesn't belong to them.

◆

It is true that Moscow in the 1990s is reminiscent of Chicago in the 1930s with gangs, mafia, shootings, prostitution, and corruption. However, Moscow's streets are still safer than, for example, New York's, and foreigners still remain

6. In modern Russia, different businesses belong to the newly established group. Since this group is very closed and their deals aren't necessarily legal, Russians call this group the financial mafia.

a privileged caste in Moscow and Russia.

In the beginning of the 1990s, when they paid a few hundred rubles for $1, Moscow was one of the cheapest places for a foreigner to live. For example, a car with a driver cost less than $30 per day; dinner in a good restaurant with Cognac and Champagne—less than $50 for two persons; a continental breakfast—less than $3; a few lovely souvenirs might be bought for less than $15; two tickets for a ballet at the Bolshoi—less than $25; and so on.

Later, when they paid almost 1,500 rubles for one dollar, all these prices above became more than twice as much. Inflation of the ruble together with the dollar became a side effect of Mr. Y. Gaidar's (together with Mr. J. Sacks, former Harvard adviser to the Russian government) monetary policies. The bad news is that Moscow has gradually become one of the most expensive cities in the world in which to live.

◆

> "Don't give idiotic advice to our politicians. If you don't want to give money, give nothing at all. We have enough of our own idiots here."
>
> (V. Tretyakov, Chief Editor, *Nezavisimaya Gazeta* daily, Moscow, December 14, 1993)

One such incongruous piece of advice was given to former Russian Vice-President Rutskoy by his American advisers. He was advised that the state should sell a certain amount of currency to different philanthropic funds at a fixed price per one dollar so that they could buy baby food or wheelchairs or medical equipment, etc. But this was Russia, and very soon this order was expanded to all kinds of chemical and oil equipment. Hundreds of contracts were signed with Western companies, but only eight deals worked. Why were not more completed? Because these Western (including American) partners suddenly disappeared along with the money. Of course they were well paid for their disappearance. Of course these previously set up funds and other entities reimbursed these rubles to the state, and, of course these rubles were worth nothing then.

> "Maybe there was something which I knew but didn't understand since some of my advice was harmful for the company. I had a Russian colleague (an emigrant) in our department and our opinions almost never coincided. If his suggestions were followed, the

company would have to stop its operations in Russia. Maybe I tried to prevent that from happening, being too self-confident. . . . He was right in repeating: this is Russia."

(Senior Adviser of a large American oil company operating in Eastern Europe and CIS; said at a conference in Michigan University's Center of Study of the USSR and East Europe, December 1993).

Rule 2–12
Try to check the origin of the money your Russian partners own, and involve them in joint operations.

Since no law system exists there, this rule is not just a warning against using dirty money. The problem is that this money may be a cover for another manipulation of the truth of which you'll never know. Or it will be too late for yóu to know.

Rule 2–13
If your American experience contradicts your Russian adviser's suggestion, follow the latter.

Most often you are trying to follow "your" common sense; Russians have their own, since different factors determine it. You may know these factors but misunderstand their importance.

◆

These factors as applied to the Russian policy toward FSU republics often seem to be almost inept if they are seen from outside of Moscow.

Let us count the lawful presidents of the countries, for example: Russia itself has two presidents[7]; Georgia, as it has been described above, has two presidents; Azerbaijan has three presidents.

President Mutalibov of Azerbaijan won the election with 92.9% of the votes, and a few months later he was overthrown and fled to Moscow. President Elchibey won 88% of the votes, and after one and a half years he was

7. A. Rutskoy was declared the Russian president by the Parliament in full accordance with the constitution.

overthrown and fled to his village. President Aliev, former KGB chief, former first secretary of Azerbaijan's Communist Party, former deputy prime minister of the FSU, won the last election in Azerbaijan with 89% of the votes. All three elections were free, and hundreds of foreigners observed.

◆

What percent of the vote do you need to be allowed to rule the country?

Apache (Denver, Colorado) had signed an agreement on some Azerbaijan offshore fields with President Mutalibov's government. Margaret Thatcher (representing BP at that time) signed another agreement with President Elchibey's government. John Major tried and almost succeeded in reconfirming the agreement above with President Aliev during his visit to London in March 1994.

Why did he *almost* succeed? Because 30% of the deal was given to the Russian Lukoil. Azerbaijan oil appeared on the world market in 1870 and had broken Standard Oil's monopoly. In 1901, Azerbaijan produced 11.5 million tons of oil; in 1965 production reached a peak of 22 million tons. And still the recoverable reserves in Azerbaijan (on and offshore) are more than 670 million tons.

The previously mentioned offshore fields (Azery, Shirak, and Gunashtly) contain as much as 50 million tons of oil (recoverable).

Lukoil is a loose association of some western Siberian producers (with headquaters in Moscow) and it had never dealt with offshore fields. Lukoil lacks money[8] and equipment, as well as experience.

You could say that Lukoil's participation is inept; but, I'd remind you that while policy seems to be the most important matter in this part of the world, personal relations are almost more important.

President G. Aliev took into account that neither the West and Americans nor Turkey or Iran were able to ensure sufficient support to Azerbaijan in its war against Armenia and in its economical needs. So he and the established powers of Azerbaijan understood that their best choice was to turn back to Russia. Not only 30% of the deal (and, possibly, the Gunashtly oil field) was given to Lukoil, but also the construction of a pipeline through Turkey, negotiated by M. Thatcher, is now postponed (forever?). BP has to use the existing sclerotic pipeline system to deliver oil to Black Sea ports or the antiquated pipeline *Druzhba*.

8. Consumer debts to Lukoi were more than 450 billion rubles ($150 million) in 1993.

How could Lukoil possibly make the deal look better? It is not possible, but it doesn't matter since this becme a political gesture. Mr. Vagit Alekperov, Lukoil's President, and President Geydar Aliev of Azerbaijan originated from the same tribe in southern Azerbaijan (i.e., they are relatives). Mr. Alekperov was deputy oil minister in Gorbachev's government and joined Mr. Yeltsin in the first days of his presidency of the Russian Socialistic Federation. Mr. Alekperov formed Lukoil with advice from President Yeltsin's circles and supplied them both rubles and currency when they were in great need. No further explanations are necessary.

◆

But something which is very curious must be added.

Two Western entities had signed a loan for Lukoil to support Russia's oil industry: the World Bank loan is $268 million and Mitcui's is $700 million.

Lukoil was going to invest this money in Tunis, Egypt, Nigeria, Trinidad and Tobago, and Azerbaijan oil industries.

Frankenburgh, a small company from the tiny country of Liechtenstein[9] is suing Lukoil for $550 million for breaking a contract on workover in western Siberia. Under the terms of this contract, the case had to be brought to an American court. But American law provides that a case against state property of countries other than the United States cannot be judged in a U.S. court. At the time when the Lukoil contract was signed (1992), Lukoil was the Russian state's entity.

They say (especially the Estonian press) that Lukoil is ready to pay some money from the loan above if some Lukoil executives will share this money with Frankenburgh.

In March 1994, Lukoil's representative explained the issue as " . . . provocated by Western competitors who intend to compromise this Russian oil company when it tries to enter the world market."

Once more, due to geographical, economical, ethnographical, and geopolitical factors, the base of Moscow's policy toward the FSU's republics cannot ever be changed. At a time when different nationalistic movements in the FSU are gaining strength, nationalists in the other republics are losing their political influence. And it is true even regarding the Baltic Republics and Ukraine. Recent events (spring 1994) show very clearly that the other republics just can't survive if totally separated from Russia; so Moldavia, Azerbaijan, and Georgia have rejoined the Commonwealth of Independent States (CIS). And some of the other republics are also looking forward to join Russia.

---

9. Frankenburgh is an American company which is registered in Liechtenstein.

In March 1994, Mr. R. Laws, representative of the American Federal Reserve System to the Byelorus government, left Minsk long before the appointed time. He said to reporters that when he was the financial advisor to the Byelorus government for 10 months, he hadn't been asked one question by the authorities.

◆

> "We have never been independent, and I cannot see any real reason why we should be independent. The only reason is the feeling of a group of nationalists who have no experience in ruling the country. We shall allow their feelings to fuse, and then we'll negotiate a reasonable economic self-rule. It would be enough for improving our industries. And, of course, we'll remain responsible for choosing our foreign partners."
>
> (V. Kuznetsov, First Deputy Chairman of the Byelorus Parliament; conversation with author, Minsk, June 1992).

In 1994 Byelorus is back in the ruble zone, using Russian currency even though Russia will lose up to $1.4 billion annually while supplying Byelorus with oil according to Russia's internal prices. Byelorus' experts would say that it is payment for having Moscow's troops in Byelorus.

In 1993 Byelorus' oil production was 2.1 million tons, which covers less than 10% of the republic's demand. Thus, Byelorus owes Russia up to 1,000 billion rubles for oil supplies in 1993, and it is a very expensive deal for Russia to bring Byelorus under Moscow's wings. Politics alone are not the only reasons to be taken into account regarding Byelorus and Moscow. The investments make economic sense, also. In Byelorus, there are three large refineries—a large gas processing plant and two rubber plants producing tires, etc. Russia badly needs this production, but the construction of new similar enterprises on its own territory might be much more expensive than supporting the Byelorus industries.

"Do Russians adhere to any religious and/or moral imperatives when doing business," the Texas businessman asked.

"Yes, some of them do. It seems these are usually less successful," I answered.

"They are patriots, as you said. Why don't these Russians which have money reinvest in Russia? They wait for us to do it. What kind of patriotism is this?"

"The reason is that your money is much more safe in Russia than a Russian's money. And, of course, your (American) money isn't safe at all."

"Aren't people thinking that their country is more important than money?"

"Yes, there are such people—those who have no money at all," I replied.

"Let me try another way," he insisted. "Is Russia a democracy?"

"I'd quote Mr. Stolypin, Russian Prime Minister in the beginning of this century: '. . . to create a democracy we need citizens, first.' It isn't my business to discuss America's problems, but I want to be more clear. Currently, some Americans are questioning the U.S. Constitution. They think that it becomes neccesary to impose more and more bans on society, for example, regarding weapons. Since you need more and more bans, it means that you have less and less citizens. Does it explain Mr. Stolypin's point of view?"

Yes, there are plenty of strange patriots, voters, and dealers in Russia.

> "Russia is great because we could conquer everybody, but nobody could conquer us."
>
> (A. Mokashov, General of the Soviet Army;[10] TV interview when running for the Presidency of the Russian Socialistic Federation, Moscow, June, 1989)

In November 1993, my friend (a priest in the Orthodox Church) and I were visiting the Moscow Cinema House for a party. A drunk man (a writer or film maker) stood upstairs watching the public enter. He was disappointed.

"There is nobody to kiss his . . . (expletive deleted)," he said to my friend.

We kept silent.

"Yes, there is no . . . to kiss."

The writer was very upset.

"So try your own," I advised.

"No profit," the man replied.

We saw a movie about the ill fortune of the Russian Tsar Boris Godunov who came to the throne a few years after Ivan the Terrible.

---

10. General Mokashov was arrested in October 1993 after he headed the attack on the Moscow TV station *Ostankino*. In February 1994, he received amnesty from the newly elected Russian Parliament.

Boris tried to conduct a kind of reasonable and liberal policy to soften the consequences of Ivan's crazy rule. Boris and his family were killed during a bloody riot in Moscow.

"What was the idea of this movie? Both of them — Tsar Boris and President Boris (Yeltsin)—failed in forcing people to lick their boots, and this is why Russia had to be destroyed," I suggested.

"In the church we keep telling our people to worship God as a ruler. We were both (church and people) punished for supporting men as rulers," explained the priest.

"This is an answer to another question, isn't it?"

"Well, we are different. Being Communist we tried hard to force other nations to follow us. Recently, they proved that they are different. Now, it is our turn."

In December 1993, my friend and I were leaving the church yard and a couple (both about 40) stood near the gate waiting to kiss the hand of their priest.

"Did you vote today," he asked.

"Yes, Father, we did."

"May I ask whom for?"

"We voted for Zjirinovsky, all of us did so, Father," answered the woman. "He will protect us, protect Russians everywhere."

"In order to protect Russians, he wants to expand Russia."

"Right, Father. Lord help us," answered the man.

"I can see you want to be a soldier."

"Oh, no, I don't, Father. God save us." he replied while crossing himself.

"Then who will do this job — Zjirinovsky himself? He promised to force you to do it."

"Who would believe anyone's promises, Father?"

After we entered the train leaving this suburb, the priest said, "Russia is lucky to have 25% of voters like these."

"Did you want to say 'only 25%,'" I asked.

"The more of them the better. . . . Nobody will cry when this guy breaks his neck, and the Savior doesn't like when people cry. I think I have understood His blessing of the poor in spirit: they are never fanatics. We never were fanatics, except once, when He allowed us to avert our faces from Him and to look into hell's abyss."

He looked out the window.

"Here we are in Moscow, once more."

"It has enough dirt," my secretary remarked. "Yesterday I finished reading Svetlana Alilueva's (Stalin's daughter) book," She thinks that Moscow is 'a volcano of human passions.' Well, it seems to be rather a chain of mud volcanoes."

"Judge not lest ye be judged," the priest replied.

Chapter 7

Saratov: The Province

In Saratov, the local operating geophysical company *Saratovneftegeofizika* (Saratov Oil Geophysical Survey) was the first to become a Joint Stock Company where the State owned only 33% of the shares. In Gorbachev's time, these games (begun in 1987) were still honest and the company prospered. It is a good example to explain the changes which took place in the FSU. Of course Gorbachev's government began the reforms in the province in order to weaken the Moscow ministries. The ministries resisted desperately, but Gorbachev used the local Party's committees and won some victories.

Up to 67% of the company's cost was given (in shares) to the personnel as a *gift*. Under these terms, however, the state would receive 33% of the net profit without any obligations for investments in or responsibility for future operations.

During the last seven years, *Saratovneftegeofizika* has modernized: 2-D seismic survey, 3-D seismic survey, precise gravity survey, electrical sounding, well seismic (VSP), geochemical survey, logging, mud logging, and core and cutting analysis.

Being free from specific restrictions (for example, to work only inside the Saratov region), the company (2,400 employees) signed a few contracts with the Kazakhstan, Kirgizstan, and Tajikistan (parts of the FSU in those times) governments and has discovered almost 80 oil and gas fields, including the supergiant Tengiz oil field in Kazakhstan.

Within the Saratov region (up to 45,000 sq km operation area), the company made more than 30,000 km of seismic lines; almost 400 sq km of 3-D seismic exploration; well seismics (VSP), almost 750 wells; logging, almost 2,750 wells; mud logging, almost 200 wells. In Kazakhstan, within an operation area of 13,500 sq km, the company made almost 15,000 km of seismic lines and more than 800 sq km of 3-D seismic exploration. In Kirgizstan and Tajikistan, within an operation area of 18,000 sq km, the company made almost 15,000 km of seismic lines; well seismics (VSP), more than 30 wells; logging, almost 700 wells; mud logging, almost 70 wells.

These figures give a good impression of the value of a mid-sized Russian geophysical company. It should be emphasized that during the last seven years, this company didn't get any supplies from Moscow, but — as a Soviet paradox — the Kirgizstan or Kazakhstan governments paid the Russian company with money received from the Oil Ministry (until the last ceased to exist). Of course, it was the not-to-be-cashed rubles, but this company — being private — could turn these rubles into cash.

Also, the company caught the best time to sell the data to foreign oil enterprises which were very active in buying all kinds of information on FSU oil fields in the late 1980s and early 1990s. This information had been sold regularly in Houston by the Oil Ministry through its entity the Central Geophysical Expedition. However, the local operating companies got less than 20% of the money earned at these sales, and, of course, they couldn't use their own discretion for spending this money.[1] *Saratovneftegeofizika* avoided these Moscow games because they had such clients as Elf Aquitaine and Chevron.

There were almost 15 other geophysical companies of the same size and shape in the FSU, and some of them were prepared to become private. But the rules changed drastically (or ceased to exist at all) after 1991, and these companies got an unclear semi-private status. Then the American advisers (to the Russian government) invented a new issue — a voucher to be changed into state enterprise shares. So the companies' personnel bought the shares of their companies, paying vouchers.

The old proverb says that the road to hell is paved with good intentions. Each Russian citizen was given one voucher (nominal cost: 10,000 rubles) in order to make the citizen a participant in the privatization of the state enterprises. These vouchers could be sold in a free market or could be turned into a certain amount of shares. The purpose of this invention was to form a market of shares and other securities. It was supposed that soon the voucher would cost much more than its nominal price, but the market price of one voucher became less than $1 (1,500 rubles) recently. So this intention to form a cash inflow to the newly privatized companies failed.

The voucher was worth nothing, is worth nothing, and probably will be worth nothing. (The last doesn't matter since the companies

1. This is why plenty of information was sold illegally.

badly need cash now.) Wild competition for the operating areas began between these companies. Each of them had to cut their personnel two or three times; some of them practically ceased to exist. The crisis is on its way, and *Saratovneftegeofizika* makes desperate attempts to survive the transition period.

In Chapter 1, we discussed a few American companies interested in data on the Fergana Valley in central Asia. The representatives of *Saratovneftegeofizika* were invited to the United States by one of them and this deal was discussed. If they were to bring their seismic field materials on the Fergana Valley to the United States for reprocessing, and if the results of reprocessing were satisfying, they would be hired by this American company for seismic exploration in Uzbekistan and Kirgizstan, if the Americans wish to appear there. Or, a job in Turkey or in Paraguay would be proposed to them,[2] if their prices were at least 20% lower than those of a Western company.

These terms looked honest to them, and they gave all the materials on the Fergana Valley to the Americans for free. This American company never appeared in central Asia or — for certain reasons — in Turkey or Paraguay. Thus, no reasons for further contacts exist.

> "I tell you this without any sadness: I learned that only one thing is honest — money."
>
> (A. Michurin, President, Joint Stock Company *Saratovneftegeofizika;* conversation with author, Saratov, November 1993).

> "I would not be amazed if I knew that they had sold our materials to another company. We should have signed a binding document with them, not just a protocol. And we should have had a lawyer, also. A word in America is worth as much as a word in Russia. I'd ask myself whether or not we have paid enough for our incompetence."
>
> (P. Zakharov, Vice-President, joint stock company *Saratovneftegeofizika;* conversation with author, Saratov, November 1933)

2. Several teams from the FSU work abroad already: *Grozneftegeofizika* works in Peru; *Byelorusneftegeofizika* works in India.

From the very beginning, keep the high reputation of your words. It will pay back much since it means a lot in Russia where this issue is a great rarity.

◆

In mid-1992, a delegation (five men and a woman) of Saratov's politicians, businessmen, and scientists was invited by a French company to visit Paris.

In the train heading to Moscow (an overnight trip), they drank 7 bottles of vodka. The next day, in the plane heading to Paris (3-hour flight), they drank 4 bottles of cognac. That night in Paris, they drank 6 more bottles. The next day in Paris they drank 34 bottles of beer while preparing for the meeting with their French hosts. So the negotiations could take place only on the third day after their arrival.

During their four days in the United States, the three representatives of *Saratovneftegeofizika* spent more than $550 for all kinds of alcoholic drinks.

◆

Russians like to drink, and then they like to keep filling in details about the drinking. This is their way to cure stresses, also. Alcoholism was and is a national problem in Russia, which might be compared with the drug problem in the United States.

Rule 2–14
Screen each potential employee for alcohol-related problems.

Your direct question could be enough since Russians don't consider this addiction as a vice. But a Russian will never admit to being a chronic alcoholic, especially if it's true. Also, make sure that the potential employee is an excellent worker during periods of boozing it up.

An American company brought a crew of drilling specialists to Saratov to work for a few weeks with Russian workers and engineers. A dipsomaniac (he was known as a brilliant specialist and had just ended his binge-drinking period in Houston — so he was included in this crew) was among these Americans, and the next day he found plenty of bosom drinking companions among the Russians. No work was done during the first week; the Russians hid this American so that he would not be sent back. After a week, the police found this man on the other side of the Volga River in a special hut; when trying to escort him to the airport, the Americans were attacked by his new *friends*, who first begged them and then forced them to release this man. And once more, he was hidden somewhere until a call from a hospital put an end to this story. It is interesting to note that the American did not know one word of Russian, and his new Russian friends did not know English.

Rule 2–15
Don't bring with you persons with alcohol-related problems. Russia is not a place for them to work.

Rule 2–16
At the table, never refuse to drink.

This is a means of establishing and maintaining personal relationships. You'd better empty your glass into your shoes or leave the party in the middle. Your absence at the table from the very beginning would be pardonable also.

◆

In the Russian culture, these deep-rooted drinking traditions were the main origin of all kinds of crime for years. From 1925 to 1990, more than 85% of the assaults were made by people who were drunk, but between 1990 and 1993 this figure dropped to less than 66%. Crime doubles annually, but more and more crimes are made consciously rather than due to drunkeness.

In the city of Saratov, crime is almost out of control. In the summer of 1993, the local authorities issued an edict forbidding children and adolescents to appear in the streets without adults after 7:00 P.M. The highways all across the region are unsafe; robbery in the inner-city trains became as common as on the street. In the city of Engels, across the Volga River, the former capital of the former Autonomous Republic of Volga's Germans, less than 200 people were killed during the past 200 years while the same number were killed in the last five.

In the middle of the 18th century, the place was a meeting point for all kinds of brigands, killers, robbers, and fugitives from Russia, Siberia, Caucasus, Persia, and central Asia. From time to time they formed large gangs. Once, under the Cossack leader E. Pugachev, who declared himself Emperor Peter III (perished husband of the Empress), their revolt threatened to dethrone the Empress Ekatherine II. The Empress had to send her famous Generalissimus A. Suvorov to defeat this huge gang. Then she decided to make the Volga River safe, and she brought more than 10,000 colonists from Germany (where the Empress herself came from). She gave these Germans the land on the eastern side of the Volga River, just across from the city of Saratov, where they established the city of Engels.

In the beginning of World War II, almost half a million Germans were deported from this area and sent to Siberia and Kazakhstan. Now they want to return and reestablish their rights to their former property along the Volga River, but, of course, they aren't welcomed there because the area has already been resettled by Russians. Neither Yeltsin nor the German governments could aid them in this pursuit, despite their verbal approval of this plan.

◆

> "Now the gangs are back and one can easily believe that this spot of land is damned. Here Russia meets Asians, Caucasians, Gypsies, and the devil himself; but the Moscow government wants us all to meet in a civilized manner. Katherine was civilized enough, but she permitted Suvorov to hang a few thousand killers and brigands. What is our civilized manner? We support private detectives and security firms."
>
> (I. Krylov, Chief of Staff of the Representative of the President of Russia in Saratov; conversation with author, Moscow, December 1993)

There are a few well-established security firms in Moscow, Saratov, western Siberia, and other parts of the FSU. These firms are headed by former police, army, KGB officers, and Afghanistan war veterans. They are equipped with machine guns, pistols, cars, armored cars, and troop carriers (is it possible in the United States?). Their personnel are well trained. They have very specific relations with both the criminal groups and the legal institutions and authorities. When they *administer justice* in their own manner, the legal institutions cover their actions.

In Moscow, three men were killed in the street near a bank office by these guards and nobody was made answerable. In Saratov, one extortioner was shot to death and another wounded by security guards, and no proceedings were taken against them. In Tyumen (western Siberia), one member of a gang was tortured until he named two of his friends, and then all three of them were killed. The local police kept silent. Sometimes the police secretly pass some specific cases to these security firms, and then they *administer justice.*

They say (and Russian mass media supports these rumors) that the Interior Affairs Ministry hires experienced killers to shoot chosen mafia godfathers since, in modern Russia, it is impossible to fight the mafia in a lawful manner. The police itself may be the source of these rumors (or is the mafia the source?).

A representative of a Norwegian shipping company told me that once he tried to explain by phone to an importunate extortioner that, a week before, a person of his kind had been killed by an order of this Norwegian company.

Be sure your shipments are well guarded when crossing Russia by rail or trucks. Be sure the storages of your equipment and your offices are well guarded. It pays to carry these large security expenses.
Be sure your guards belong to a legally established security firm.

Rule 2–17
Avoid using the police either as security guards or as detectives. Instead, try to enlist the services of private companies to provide these functions.

The criminal groups frequently have special informers in the police. Those private companies, however, have their own mechanism of keeping the gangs at a distance by intimidation and clear consequences of violating the code.

Rule 2–18
Avoid conducting any financial operations in cash, either rubles or currency.

These criminal groups try to have special informers everywhere, including at banks.

◆

Two petroleum basins (Volga-Ural and Pre-Caspian) adjoin one another in the southeastern corner of the Saratov region. The first one as the part of the Russian Platform is shallow (2.2–2.9 km), and the Devonian and Carboniferous deposits are oil and gas saturated; almost 30% of the initial in-place reserves of this area were produced during the last 60 years. The second one eastward and southward gradually becomes as deep as 15,000 meters or more; the thickness of Permian and Triassic deposits rises up in these directions, and the salt beds and salt domes appear in the geological sections. The estimated oil reserves within the Pre-Caspian Depression are as much as 28 billion tons, and less than 25% of them are located at depths less than 7,000 m.; of these only one-tenth has been discovered.

In the Saratov region, the northern edge of the Pre-Caspian Depression has not been explored by drilling, and therefore, its perspectives are unclear. Along this edge, a few hundred kilometers to the southeast in Kazakhstan, a gigantic oil-gas-condensate Karachaganack field was discovered, and British Gas is conducting the development of this field.

◆

> "I understand what the French oilmen are looking for there: another Karachaganack. I didn't recommend my company to fight for the neighboring area in the Saratov region since I had some geological doubts. Then I saw the data on Tengiz (Kazakhstan) and Yuzno-Plodovitenskoe (Kalmykia) located along the edge of the Pre-Caspian Depression and I realized I was wrong. This is Russia and everything is possible here. The reserves of oil are enormous, and their concentration within a limited area reminds me of the Persian Gulf. I've learned my lesson, and I am thinking of returning to Saratov again. It's a pity it is not on a coast."
>
> (Vice-President, Exploration and Geology, of a large American oil company; conversation with author, Houston, January 1994)

A French oil company started negotiations to explore in the Saratov region with the FSU in 1991. Then they negotiated with the Russian government and were told to discuss this idea with the local authorities (while the Russian government was playing games of decentralization). The local authorities didn't find appropriate laws to make decisions by themselves, and when they finally did, the Russian Supreme Soviet[3] played against Yeltsin by advocating centralization — and they decided to study the idea first. It took

3. The Russian Supreme Soviet (first Russian Parliament) was elected within the FSU, and Mr. Yeltsin was its first chairman. Then the FSU ceased to exist in 1991. This parliament remained active until October 4, 1993.

them more than a year to study the French proposal in different commissions and sub-commissions until all of them disappeared in October 1993 (when Yeltsin crushed the Russian Parliament uprising). Then the Russian prime minister tried to play centralization once more, and Saratov's local authorities waited.

> "Yes, it is a shame. . . . It seems that the French should have been brought to Saratov by Napoleon. At least, they would know when to leave Russia. But if you want me to be serious: this is the fortune of our province."
>
> (P. Zakharov, Vice-President, *Saratovneftegeofizika;* conversation with author, Moscow, December 1993)

I'd repeat that nobody wants to give up control of oil to the foreigners. However, the local authorities are compelled to accept it since they badly need money for developing their regions. But the Moscow government would be sensitive only to the needs of the most important provinces (western Siberia or Komi, for example) or to the needs of the national republics (Tatarstan, etc.). Since provinces like the Saratov region have no other choice but to wait patiently, the Moscow government is trying to maintain the status-quo. In order to establish and conduct business in these provinces, you'd have to obtain as allies the local authorities and the old oil aristocracy (local and Moscow's).

A small American company had succeeded in getting an exploration and developing project on the western side of the Volga River by helping a small Saratov company to obtain a few small turboexpanders for producing gasoline from condensate. The president of this Russian company was the son of the chief geologist of the local oil producing company (owned by the state, of course).

Once more, personal interest is a powerful factor in the Russian provinces also.

Chapter 8

Calculations on Western Siberia

The western Siberian region is as large as 3.5 million sq km, and the area of the most prospective and promising oil and gas saturated zones is almost 2 million sq km (for oil saturated zones — 0.9 million sq km). Within the discovered oil fields, the average porosity of Mesozoic sandstones is about 12.5%; their average oil saturated thickness is almost 6.5 meters; their average oil saturation factor is almost 80%. The product after multiplying these figures defines the estimated potential original oil (in place) as great as 500 trillion metric tons. (The average specific gravity of Siberian oil is 0.88 g/cm^3.) The average permeability of the reservoir rocks is up to 340 mD; the depth — less than 2,500 meters; the initial production rate — more than 190 metric tons per day; the initial bottom-hole pressure — up to 260 atm.

As for now, only the fields along the western Siberian rivers are explored and developed. Within the western Siberian petroleum basin, it is almost impossible to find a spot where the Mesozoic deposits aren't oil (gas) saturated.

This western Siberian basin is unique. It is well known that the main oil (gas) basins all over the world are located within different foredeeps, inter-mountain depressions, or within those platform zones where the oil (gas) saturated deposits are expanded toward foredeeps or surrounding highlands. Within the western Siberian basin, the Mesozoic deposits pinch out toward its edges in all directions and cease to exist outside this basin.

The western Siberian sedimentary basin covers a typical tectonic plate. The western edge of this plate exposes at the surface approximately along the meridians of 62 to 64 degrees of longitude and the eastern edge along 80 to 82 degrees of longitude. Northward the plate dives under the waters of the Arctic Ocean. The average thickness of the Mesozoic sediments which cover the plate is almost

2,500 meters; the maximal thickness is up to 4,860 meters and may be more in some unexplored local depressions. The Jurassic sequences are oil saturated in the southern half of the basin; the Cretaceous sequences are gas saturated in the northern half of the basin.

The first gas field (Berezovo) was discovered in 1953; the first oil field (Shaim) in 1960. Since then, as many as 43 gas fields and 34 oil fields were discovered, and more than 4 billion tons of oil were produced.

The western Siberian basin produces on average 60% of Russia's overall production. The peak of almost 400 million tons was reached in 1988; in 1991 the basin produced less than 320 million tons; in 1992 it produced slightly more than 200 million tons from almost 19,000 wells.

All the figures on the decline of production are statistics which are given to the government. Are the real figures available to the government? It is doubtful because too many people are interested in concealing the true amounts of oil they deal with (produce or transport or sell).

> "Just to survive, I would not report to them (the government) the whole amount of oil I produce. I must feed my workers, local authorities, clerks, and higher ups who help me to find approval for my false reports in Moscow — pipeline owners all across Russia, port authorities at the Black Sea and Baltic Sea, and God knows who else. Of course, the decline of production is large enough, isn't it?"
>
> (Chief geologist of a large Siberian oil production company; conversation with author, Tyumen, December 1993)

The reasons for the real decline are not only geological; the numerous other reasons are rather technical, economical, and political:

- The decentralization of the oil industry meets desperate resistance in Moscow.
- The achieved level of decentralization damaged the possibility of state investments in exploration and development of the new oil fields located at the far corners of the basin.

- The oil industry was and is very inefficient and chronically underfunded since the oil was absurdly underpriced. (In 1991 gasoline was cheaper than mineral water, for example.)
- The equipment is old and in a state of disrepair.
- The delivery of new supplies has been impeded by the break up of the FSU.
- Russia's macro-economic problems have precluded large-scale imports on commercial terms.

The western Siberian basin (principally the Urengoy and Yamburg gas fields) holds 35% of the world's gas reserves (approximately 50 trillion cubic meters). In 1991 more than 800 billion cubic meters of gas were produced in the basin. There has been little change in production levels over the past five years. Approximately 100 billion cubic meters is exported annually, and this represents the single largest source of export earnings for Russia. Seventy-five billion cubic meters of gas are used as compressor fuel and ten billion cubic meters is flared in place; the remainder is consumed across the former Soviet Union (FSU).

Gas production has not slumped like oil production, which is largely due to the fact that it is a much younger industry, and therefore most of the equipment is still managing to function. (The technology is also less sophisticated than that needed in the oil industry.) The Russian government controls the gas industry, and, even though the state's investments have dropped 50% in the past five years, they remain at a satisfactory level due to the personal care of Russian Prime Minister Chernomyrdin. (Mr. Chernomyrdin was the Minister of the Gas Industry in President Gorbachev's government.) But still the leakages from aging pipelines are up to 20%.

In Bashkirya, on a summer night of 1989, two trains met at a small station where the railway crosses a natural gas pipeline. Suddenly, both trains were enveloped in flames from a gas flare explosion. More than 200 people perished in the fire.

The poor-quality Russian compressors and insufficient and aging Western compressors aren't able to cope with their tasks; insufficient production capacity upstream and (linked to this) the inability to store gas for peak requirements make the losses of gas enormous.

There is no refinery within the described region.[1]

---

1. Eastward, following the Trans-Siberian Railway, there are large refineries in Omsk, Tomsk, and Achinsk. The location of the refineries outside of the described oil producing region has an exclusively political meaning: to keep it dependent on Moscow.

In order to try and boost production, attempts have been made by the different Russian governments to allow greater autonomy to the local operating companies and local authorities in western Siberia.[2] Under a presidential edict, the regional authorities were, in theory, entitled to proceeds from the sale of 10% of the regional production, and the local companies and their associations were entitled to sell 10% of their own production (which was increased to 40% in a subsequent resolution). However, in reality this was happening in a short time only. Oil production fell to such an extent that state quotas for oil (which were to be phased out gradually by 1993 and 1994) could not be met; under a later government decree, the region has been forced to meet these before they are allowed to release their sales rights.

From 1991 to 1993, even at the increased state purchasing prices, sales to the government rather than on the free market were unattractive to local industry. Some semi-legal and illegal ways were chosen to sell the oil on the free market. Millions of dollars were made by those individuals who succeeded, but these dollars were lost to the industry. Meanwhile, the cost of equipment has spiralled way out of proportion to the recent increases in the oil prices, and a vicious circle has developed. The Western Siberian industry is critically short of the spare parts needed to boost oil production (between 1991 and 1993 they received only 35% of what they needed), and they were prevented from earning money with which to buy them. Therefore, more oil wells became idle, production plummeted further, and earning continued to fall.

Again, a new issue has arisen — the crisis of non-payment. The monetarists of Y. Gaidar's government invented a situation where almost each Russian enterprise became short of money; therefore, the consumers of oil had no means of paying for it. (Everybody in the FSU has become cautious about barter deals.) Although oil prices were brought up to world levels (and above it, in some cases) at the end of 1993, it didn't affect the local oil industry since the consumers didn't have money to pay for it.

The Siberian oilmen didn't and don't see their large (on the Russian scale) salaries for months on end, and inflation eats up a large chunk of the salary each week.

2. The Autonomous Republic of Khanty and Mancy and Yamal Nenetz Autonomous District are located within the described region.

> "Almost half a year ago, I received my salary for the last time. I received 680 thousand rubles, which was worth $618 then. If I had received it in time, it would have been worth $755. Now, in our bookkeeper's office, I see the sum they owe me, and it is more than 4 million rubles. If I had had this money in time, I would have received $3,300 dollars; but today it is less than $2,850 and still I don't have it. These $450 are a great sum for us; I with my family spend less than $200 each month."
>
> (V. Belyakov, Senior Geologist, *Samotlorneft;* conversation with author, Tyumen, December 1993)

The pressure from the oil regions for greater economic autonomy has been increasing, and they are undoubtedly growing in strength. They are seeking foreign investors independently, and they might become reliable partners, even though Moscow is trying to win the initiative.

A few smart moves were made in Moscow when Mr. Y. Shafranik, a famous western Siberian leader, was appointed as the Minister of Fuel and Energy of Russia. And, also, the government tried to clearly demarcate the main areas where it would support foreign investments.

◆

Naturally, the Russian government is eager to attract credits and financial support for purchases of equipment. The government continues to be active in attracting major aid — possibly from the World Bank, Export-Import Bank, and others — to rehabilitate wells in western Siberia and to develop discovered fields located at the far corners of the region. It is also understandable that those banks aren't as easily attracted by these plans.

The government's attitude towards investment by Western oil companies, however, is more cautious. Currently, there are few operational production schemes in western Siberia involving joint investment. In each case, production is said to have slowed year by year because of fiscal measures that have reduced profitability for the Western partners. (They say that the first joint ventures in the described region — White Nights and Polar Lights — both have declared similar annual losses of up to $80 million in 1993.)

The indications are that the regional authorities and the local oil enterprises remain cautious about the implications of Western equity involvement. Nevertheless, you can be assured that in western Siberia, due to its geographical location in the middle of Russia, the local opposition isn't as fierce as at the country's edges (Sakhalin, for example).

Some foreign investments in the development and improvement of the oil industry in western Siberia are welcomed by the Russian government, local authorities, and local operating oil companies. The following five items are the main investment directions which will obtain clear support:

◆

1. Secondary recovery and well repairs at the producing fields:

   Each of the previously mentioned producing fields still contains from 70% to 80% of initial reserves (in place). In total, more than 2,800 wells within these fields don't produce oil due to different reasons. A few American and Canadian oil companies are working in the region now, and a portion of the increased production of oil belongs to them. Some of the companies have found the deal profitable — some have not.

   > "We have lost not only our investments, our equipment,[3] our hope — we have lost the spirit. We feel trapped here. Horrible conditions, stupid laws, bad workers — what is good? Nothing."
   >
   > (Manager of a Canadian oil company; conversation with author, Tyumen, December 1993)

   > "I would never come here to work if I had my salary in Calgary. In the summer you aren't permitted to breathe because of the mosquitoes; in the winter you can't breathe because of the frost. It looks like Alaska but it is much worse. I used to work every day in Alaska; here, I work one day a week. Why? Today, there is no mud for drilling; tomorrow, my Russian partners are drunk; the next day, the well logging crew doesn't appear."

   > (Technician of the same Canadian company; conversation with author, Tyumen, December 1993).

3. Nobody would take back equipment since the shipment may cost more than the equipment itself.

> "This year we have receipts which will bring a third of our invested money back. I hope we will make some net profit next year. It seems it will excuse our presence here. But they can easily damage the deal if they raise the cost of pumping the oil across Russia once again. What happens then? Then we leave."
>
> (Director, Russian-American joint venture; conversation with author onboard a Delta flight, November, 1993)

For many cases in western Siberia, rigs, pumps, cement, and other straightforward equipment are required rather than other sophisticated techniques of secondary recovery.

2. Exploration, exploratory drilling, and development of oil fields and promising areas located at a distance from the main rivers and devoid of any logistics:

   The Russian prime minister had approved a list of such fields (five located in the Tyumen region) in the beginning of 1993, promising certain tax holidays for the investors.

In the summer of 1993, a CEO of an American oil company was lured to see one such place at the southern corner of central Siberia (Irkutsk region). He flew there with a vice-president of a Moscow-based private company, their geologist, and the mayor of the small city of Ust-Kut, located *only* 350 miles from the field. They spent a day flying from Moscow to the city of Bratsk. The next day they spent in a helicopter flying eastward to the mayor's city. The third day the geologist was drunk, and they spent their time in a poor hotel where each room housed six people and billions of mosquitoes. The weather was bad the next day, but in return they had a good dinner with the mayor, his family, and friends accompanied by Russian songs and Siberian dances. The dinner began at 2:00 P.M. and ended (for the Americans) at midnight. The hosts continued to drink, eat, and sing, and the Americans could hardly find their way back to the hotel through the dark unpaved streets of Ust-Kut. Finally, on the fifth day, they visited the field. Their helicopter landed at a knoll among the marshes, at a distance of one-half kilometer from a rig which they couldn't approach since the rig was surrounded by a lake . . . of crude oil produced from the well.

"There was no oil here in the winter," the geologist screamed. "The casing didn't appear to be damaged."

"Perhaps the cement ring is torn somewhere and the oil leaks," the foreman answered.

"What sort of cement were you using?" the American asked.

"Whatever was delivered. It isn't easy to bring supplies here," was the answer.

As they flew around the rig, the American made an observation.

"Can you see that the rig is lopsided?"

"Summer — the soil is melting . . . it doesn't matter; we will pull it to another spot this winter."

Two other wells which they visited seemed to be in order.

Next day they traveled four hours to Irkutsk, but they missed the flight to Moscow and enjoyed another night in Siberia. The hotel was much better; a few mosquitoes were easy to kill; but the food in the restaurant was bad.

The next day, after the Americans returned to Moscow, I bet them $100 that I could obtain in Moscow the same data on the field they were shown in this far corner. (Russians call such sites *the bear's corner.*) Of course I won their $100.

All the data and information you may need is available in Moscow.

The *bear's corner* above is a good example to explain in detail why the government welcomes only foreign investments for certain projects (the Governmental List) and why foreign investors probably ignore this invitation.

As a matter of fact, three oil fields (Markovsky, Yarakhtinsky, and V. Chonsky) were discovered in this *bear's corner* visited by the American CEO in the summer of 1993. A study of these fields shows the following average figures:

- Depth of occurrence: 2,450 m
- Thickness of oil-bearing rocks: 40 m
- Thickness of oil saturated rocks within active zones: 28 m
- Porosity factor: 16%
- Reservoir temperature: 36 degrees (Celsius)
- Reservoir oil density: 0.77 kg/cu m
- Degassed oil density: 0.826 kg/cu m
- Gas factor: 210 cu m/cu m
- Resins content: 4%
- Asphaltenes: 0%

- Paraffins: 1.8%
- Sulphur: 0.43%
- Saturation pressure: 20.1 mPa
- Number of explored producing wells within the contour: 11 (at the total 3 fields)
- Commercial oil reserves (each field): 23 million tons
- Recoverable reserves (each field): 6.5 million tons.

Let's look at some rough but simple calculations of the cost of developing these fields:[4]

Expenses related to oil well drilling[5]

Initial data:

- Average depth of 1 well: 2,600 m
- Estimated cost/well: $0.2 million
- Production rate/well: 60 tons/day
- Total annual production/well: 20,000 tons of oil
- Number of producing wells: 250
- Cost of drilling 250 wells: $50 million

Expenses based on the projected annual oil production and number of wells by years:

- First year (30,000 tons/2 wells): $0.4 million
- Second year (100,000 tons/3 wells): $0.6 million
- Third year (700,000 tons/28 wells): $5.6 million
- Fourth year (5,000,000 tons/217 wells): $43.4 million
  Total cost of drilling for 4 years: $50 million

Cost of the pipeline construction[6]

- Yarakhta/Markovo/Ust-Kut pipeline (500 mm diameter) — $36 million
- Yarakhta/V. Chona pipeline (100 mm diameter) — $18 million
- Infrastructure at the Markovo field: $1.56 million
- Infrastructure at the Yarakhtinsky field: $4.5 million
- Infrastructure at the V. Chonsky field: $10 million
- Acquisition of motor transport means: $1 million

---

4. These figures are taken from a pre-Feasibility Study made by the author for an American oil company.

5. Calculations are based on the projected rate of oil production of 5 million tons per year.

6. These costs are for connecting these oil fields with the railway.

Total cost of the project: $138.46 million plus some additional unforeseen expenditures (10%) equals $153 million.

You need plenty of imagination to calculate the income, but let us try. Suppose we have a ready buyer for the oil on the Siberian railway station of Ust-Kut, and let us suppose he is paying 100 dollars/ton of crude oil (approximately $15/bbl).

Cost of oil to be sold in Ust-Kut:

- First year (30,000 tons): $3 million
- Second year (100,000 tons): $10 million
- Third year (700,000 tons): $70 million
- Fourth year (5,000,000 tons): $500 million

Net receipt (Gross receipt less taxes, operating costs, and costs of transportation to Ust-Kut) shall make up:

- First year: $2.1 million
- Second year: $6.5 million
- Third year: $42 million
- Fourth year: $275 million

Total net receipt for four years shall make up: $325.6 million

Still it looks nice, but it is you who stands at this railway station with money to pay for delivery of your oil at a distance of 3,000 miles to a port in the Russian Far East; and even though only half of the produced oil is your portion, you still need 100 railway tankers for daily transportation. Then you shall pay for shipment of your oil to Japan or South Korea.

Rule 2–19
Don't hunt for fields located in the far corners *(bear's corners)*.

You cannot predict the problems to be faced in this type of venture.

> "There you begin from the very beginning of human history — from inventing a fire for lighting the night. And there is nobody to begin the history with since no human being appears for hundreds of miles around. It is really enjoyable to stay away from there. By the way, I had never seen oil lakes before."
>
> (CEO of an American oil company; said at dinner after returning from Irkutsk; Moscow, June 1993)

3. The third acceptable investment direction is improving (as much as is possible) the ecological disaster which has resulted from the oil industry's activity within the described region.

Flying over western Siberia, you can see huge black spots within the taiga and marshes. These are oil lakes formed in places where the pipelines leaked. In Tyumen, the authorities of the state company *Sibnefteprovody* (West Siberian Oil Pipelines) report that almost 2.9 million tons of oil have been lost during the last 10 years; in Moscow, the sources in the Ministry of Ecology report losses were up to 4.4 million tons during the last 5 years.

Also, during exploration drilling within the fields, large amounts of oil were lost in the process of testing the wells.

And, the local operating companies — in order to achieve the planned (in Moscow) level of production but having no ready infrastructure (pipelines, pumps, etc.) — stored the produced oil in barns (holes dug by bulldozers). This oil was produced and paid off, but it remains in these barns, covered with soil by the same bulldozers.

In 1992, a tiny Austrian company formed a joint venture with an entity of a local operating oil company in western Siberia. The purpose of this venture was to clean the barns. In the same year, this joint venture reportedly produced 0.76 million tons of oil and put it in the pipeline. According to the Austrians, they received up to 340,000 tons of crude oil in the Black Sea port of Novorossisk and then sold it at market. During the first half of 1993, they reportedly produced 0.2 million tons and received 65,000 tons of crude oil in Novorossisk. Then the Austrians' partners disappeared and a few of their Russian partners were arrested. The state investigating bodies ascertained that the oil produced by this joint venture existed . . . on paper only. In reality, not one barrel of oil was produced since none of the barns were cleaned up. But the paperwork was done promptly and properly.

A similar case was brought to the court in Volgograd (former Stalingrad) early in 1993. A Finland-Russian joint venture which was registered there managed to clean barns in western Siberia, also. They recovered almost 6,000 tons of heavy oil from a barn but in the documents the figure of 56,000 tons was reported.

Rule 2–20

Check the figures which your Russian partners report in the documents. They will try to exaggerate your achievements. Advise them to be modest.

4. Preventing pollution caused by flared and unflared associated wet and natural gases is another acceptable investment area.

   In western Siberia, less than 60% of associated gas has been utilized (and this is well above that in other Russian basins).

   The local hunters have adapted very well to this situation. A new method of hunting capercaillies (grouse) and heath cocks was invented in the taiga around the oil (gas) fields. Some bird food is attached to the gassing christmas trees (none of them is ever sealed) and after having their dinner, the gassed birds become an easy bag for hunters. I was told that this meat tastes bad.

5. Construction of a few small refineries to supply local machinery with fuel (diesel and gasoline) is acceptable also. The residue of hydrocarbons (after extracting the products above) could be pumped back into the pipeline.

   The idea of constructing big refineries will not find support in Moscow for purely political reasons. The more crude oil used in-place, the less crude would be transported westward where the huge Russian refineries are located. This would mean the appearance of hundreds of thousands of jobless people just in the center of European Russia and, also, further loss of central control over the oil industry.

> "I would add a few economical reasons, also. In western Siberia, there aren't enough consumers for fuel and the manpower to serve at the large refineries."
>
> (E. Taslitsky, Deputy Minister of Fuel and Energy; conversation with author, Moscow, June 1993)

It is cheaper to transport products than crude oil. And, more than half of the Siberian enterprises stand idle; therefore plenty of manpower exists there.

◆

The company *Sibnefteprovod* (Siberian Oil Pipelines) is collecting all the oil produced in Siberia and transporting it in four directions:

- Toward the refineries located in the center of European Russia (Nizhny Novgorod, Moscow, Yaroslavl, Ryazan, etc.).

- Toward the refineries located in the Volga-Ural province (Ufa, Ishimbay, Samara, Syzran, Saratov, etc.) where the sweet Siberian oil is mixed with local sulphurous crude to reduce the cost of refining the last. The famous *Soviet export blend* is made also and shipped toward the Druzhba export lines (long-distance pipelines) and Black Sea ports.
- Eastward to the refinery in Omsk and Tomsk.
- Southeast to the refinery in Omsk. A special pipeline for liquified gas (LPG) collected across the basin's oil (gas) fields was built toward the city of Tobolsk where a LPG fractionation facility is located. This pipeline belongs to and is operated by *Sibnefteprovod* also.

◆

The total length of the company's pipelines is up to 11,000 km. The company transports 600,000 tons of oil per day. The total capacity of its pipes and storage tanks is up to 2.5 million cubic meters; the capacity of the storage tanks is up to 1.4 million cubic meters. These storage tanks are located in 23 spots where the oil produced by the local companies leaves their own collecting systems for long-distance transportation.

The vapor recovery facilities exist neither at the producing companies' storage tanks nor at those which belong to *Sibnefteprovod.* The air pollution around these storage tanks makes it necessary for workers to wear masks, especially in summer time; the annual losses of hydrocarbons are equal to 3% of the oil pumped through the *Sibnefteprovod*'s pipes.[7]

The company had paid a fine equal to the cost of more than 6 million tons of oil (average annual losses) to the Ecological Ministry. As long as *Sibnefteprovod* remained a state entity, this didn't bother the company's leadership too much. In early 1994, the company was transformed into a joint-stock corporation, and they saw that there was a better way to use this huge amount of money. They looked for a foreign partner for the vapor recovery project, especially after the successful start of such an undertaking by Viaperking (a subsidiary of Global Natural Resources, Inc.) in Tatarstan.

The potential of this project looked nice and clear since the recovered hydrocarbons have a ready market in Tobolsk. The local chemical plant had an agreement with consumers in Finland and was paid for LPG in currency. The plant owned a few hundred railway tanks for transporting the product to Finland.

---

7. This figure was given by Mr. Chepursky, chief engineer of *Sibnefteprovod.*

Mr. V. Chepursky, chief engineer of *Sibnefteprovod*'s headquarters in Tyumen, dialed Tobolsk's number, pushed the speaker button, and asked the general director of the chemical plant:

"How much extra LPG would you buy from me in addition to what you have now?"

"Unlimited. The more the better. And I will give you a good price, Vladimir."

We could hear the answer. A few representatives of an American corporation were sitting in his office.

"Besides," continued Mr. Chepursky, "we have plenty of consumers for the household gas and other products. Of course, these will be paid in rubles. Okay, we accept rubles since it is a sure gain for us; you'd receive the currency. Besides, if we declare our joint venture as an ecological enterprise, we will enjoy a long and large tax holiday. Do we have a deal?"

"Yes, we do," the Americans answered.

The same day, an Application (Memorandum of Association) was signed by both sides.

◆

Four documents have to be submitted in order to register a joint venture in Russia:

1. Application (Memorandum of Association)
2. Charter of the Joint Venture
3. Protocol of Intentions to form a Joint Venture
4. Feasibility Study

Only the first two are binding documents.

Joint ventures related to the oil (gas) industry are to be registered in Moscow, others with the local authorities.

Rule 2–21

Hire a Russian lawyer to prepare all the documents needed to register your joint venture.

Do it by yourself (hiring the lawyer). You can be sure that your Russian partners will not have done this job properly. Also, they might try to bribe clerks secretly.

The Protocol of Intentions on the vapor recovery project above was signed in the first week of November 1993. It stated that a private Moscow company (the third partner which brought these

Americans to Tyumen) is responsible for having a lawyer prepare all the necessary documents. Of course, those documents were not ready until the lawyer which the Americans had hired made the proper job. And it was finished the night before Christmas; then this meeting with Mr. Chepursky took place.

In the beginning of 1994, this American corporation faced further financial problems at home. (It was in a very shaky position before, also). This corporation was formed by several small companies specifically for the purpose of selling various remanufactured oil and gas hardware to Russia. When forming the corporation, no money was invested by the participants — only different kinds of used equipment. Some of those companies had a history of bankruptcy and were burdened with debts.

Rule 2–22
Don't seek to solve your domestic financial problems in Russia.

Working in Russia, you will need solid reserves of time and money to eventually make a profit.

Of course the corporation was concealing its financial weakness from the very beginning of the negotiations with *Sibnefteprovod* (and succeeded since this matter is difficult to recognize even inside the United States). Then, after a binding document was signed in Tyumen, the American partners disappeared without any explanations. Probably they were looking for a bank loan or credit or another partner. But the deal was hung up; other foreign firms didn't want to interfere since a binding agreement had been signed by *Sibnefteprovod*.

Some harmful consequences of promoter involvement is shown in the chapters on central Asia.

> "I expressed to (former) President Nixon our extreme concern about the growing anti-American mood of the Russian public. Such feeling had not been seen in the darkest days of the Cold War."
>
> (G. Zuganov, Chairman of the Russian Communist Party; said in an interview after a conversation with former President Nixon, Moscow, March1994)

"What we Russians didn't expect is that some Americans are as thievish, furtive, and stealthy as we are."

(M. Tolstoy, Ph.D., member of the former Russian Parliament and of state Duma; conversation with author, Moscow, December 1993)

"We Russians are great idealists. We expect close adherence to the ethical and moral laws from other people. This is because we ourselves break them so easily. I am sad to note that for Americans, in their country also, religion is no longer a guard against the evil that men do."

(Priest of the Orthodox Church; conversation with author, Moscow, November 1993)

"There is only one way to stop these jokes: we must become competent or these hairy mongrels[8] may kill the great possibilities which appear occasionally in Russia and might just as easily be reversed."

(V. N. Chepursky, Chief Engineer, *Sibnefteprovod;* phone conversation with author, March 1994)

A small thing should have been inserted into this agreement above — the date of its expiration. But no governmental resolution ever required this feature. All across the former FSU, these rules of the game (no third-party interference) are unknown, and this is why hundreds of projects remain on paper.

◆

No legislation on foreign investments, property, or taxes ever existed in the FSU. It took a long time for FSU bureaucrats to prepare a contract, but in return one could be sure that each item of the signed contract would be observed.

No legislation (as a system) on the matters above exists in modern Russia, either. It is very easy to have a contract signed, but you can be sure that nobody is obliged to follow any agreements. The Soviet Parliament, and later the Russian Parliament, tried to change this situation, but both of them were too involved in power struggles to pay much attention to this issue.

8. In Russia a contumely outrage.

As of today, only several laws are on the books, but they are virtually unenforceable because of dozens of further governmental resolutions, a large portion of which are secret (the old Russian way to corrupt the laws).

A law on mineral resources had been adopted in early 1992. This law outlined the procedures for the exploration of Russia's natural resources, and for the first time codified the use of concessions and production-sharing arrangements in the Russian oil and gas industry. However, government sources have indicated hundreds of times that although concessions will be made to foreign companies, they are likely to be limited both in scope and in number. A resolution was passed by the government on the exact licensing procedure, which meant that a separate law on licensing was unlikely.

Another law — on environmental protection — is of considerable relevance to oil and gas companies because it provides details of liability for any activities which have a detrimental effect upon the environment. Also, this law provides certain tax holidays for ecological enterprises but leaves the determination of exact figures for the government. It means that enterprises involved in ecological activities aren't getting any tax holidays automatically but only after further contacts with bureaucracy, the results of which are always unpredictable.

Another example is the mineral resources law which specifically keeps gold and silver within the government's purview. However, a few weeks after the law was adopted, a secret governmental resolution declared the state's monopoly over all the gold deposits and the recovered gold in the entire country. In the beginning of March 1994, President Yeltsin's decree abrogated both the law and the resolution above.

Under Yeltsin's decree, the state's quota is *only* 50% of the recovered gold. It means that producing companies are obliged to sell 50% of recovered gold to the state at fixed prices. The rest is to be sold to the Central Bank and five to eight commercial banks through different auctions where currency is permitted to be used. The foreign companies are permitted to participate in gold recovery, and then they may take their share of gold to the world markets.

However, it is very unclear how the state could pay for its quota since its debts to the producing companies were up to 614.7 billion rubles and $210 million[9] at the end of 1993.

Of course, the president's decree stated nothing on taxation, even though this issue is a major source of concern to Western investors.

The problem of taxes in the FSU cannot be discussed in any print media since the regulations are changing unexpectedly and almost monthly.

◆

9. A. Golovaty, Deputy Minister of Finance, announced these issues at a press conference held at the end of February 1994.

> "They (Amoco) are doing business under rapidly changing conditions. It's been almost a daily matter of what your tax structure is going to be."
>
> (J. Harvey, analyst, DLJ Securities; interview with U.S. newspaper reporter, September 1993)

In the beginning of 1994, Russian taxes on oil became (in total) almost $70 per metric ton ($10/bbl) and retroactively damaged the profits of joint ventures formed a few years ago.

Visiting Moscow in March 1994, U.S. Secretary of Commerce R. H. Brown evidently was unable to get any promises of relief on the export tax. As an alternative to excise taxes, the U.S. government proposed royalty agreements on sharing production. It seems that the Russian government understood that this provides some currency to Russia, and it was declared that a controversial $5/bbl export tax on oil is to be reviewed.

In the beginning of April 1994, President Yeltsin's decree stated that within three years, no new taxes can be introduced and that no tax rates can be increased and that the three-year tax holidays for joint ventures cannot be changed. But taxpayers know that further interpretations will lose the primary sense of this decree. When I asked Mr. V. Udod, Deputy Director of the Regional State Tax Management branch in Tyumen, to comment on the president's last decree, he answered: "No comment since no written details were given to us. . . . Believe me, nothing has changed."

I'd repeat once more that there are hundreds of ways to avoid paying any taxes or to pay reasonable taxes if you follow your Russian partners. None of these ways are in direct violation of laws if the appropriate bureaucrats would use the correct methods of approach to the matter.

> "Cooperation with the USA is a priority for our country. . . . If the Congress of the USA favors massive investments in Russia, the Duma will try to create auspicious conditions for American businessmen, at least for a few years."
>
> (I. Ribkin, Chairman of the State Duma;[10] said at a meeting with U.S. congressmen, Washington, March 8, 1994)

10. Mr. Ribkin is a member of the Russian Communist Party.

"The Soviet Parliament consisted of 3,000 members; the previous Russian Parliament — of 2,000; the last one — of 1,000. Each member of each chamber had been invited to the U.S. and has been promising to change something in Russia. It seems they didn't have enough time to fulfill their promises. Aren't we paying too much money for words?"

(U.S. Congressman; phone conversation with author, New York, February 1994)

Now, let us have a quick look at some long-term issues in western Siberia.

For the moment the Russian authorities couldn't (even with well-targeted and guaranteed Western investment) develop gas and oil fields in relatively inaccessible parts of the region. But the situation will need to be changed within the next 5 to 10 years, when new trunk gas pipelines will be needed. Also, within the same 10 years, major new exploration may be needed in the eastern and northern parts of this basin. The economics of hydrocarbon production may also justify exploration offshore in the Barents Sea (where a few gigantic fields have been already discovered) and in the unexplored Kara Sea. This is an area where Western technology and engineering skills may in due course come into play, although the problems arising from pack-ice and arctic conditions still pose problems that cannot be solved readily at the present level of technology.

The Russian scientists' idea of using idle submarines for exploration and drilling from the Arctic Ocean's bottom doesn't look crazy, even today, since the costs of disassembling or conserving them might be equal to those of their re-equipment.

Western Siberia seems to be less dependent on political changes in Moscow since any central government would depend on oil money. If V. Chernomyrdin is the next Russian president,[11] he will try to attract foreign investments by issuing a kind of governmental insurance for them. Then, of course, his government would choose and approve partners for Siberian enterprises. This will cut the opportunities for small and mid-sized Western companies to enter, but those with a good reputation within the region will enjoy more security.

11. In April 1994, the International Monetary Fund approved a $1.5 billion loan for Mr. Chernomyrdin's government, which was a significant moral support for its policy even though this isn't enough money for Russia's needs.

If the nationalists win, their anti-Americanism will be restricted during the first few years both by the united (it would be dangerous for them to remain disunited) opposition and limited financial sources. Also, there is nothing they could set off against the fully convertible ruble. And if the ruble is convertible, it means that the oil world market is gradually approaching western Siberia which will prevent the risks of overdependence on the Mideast oil producers.

Chapter 9

Black Sea: Onshore and Offshore

Early in the 15th century, Sofia Paleolog, the daughter of the last Byzantine Emperor, married Moscow's Grand Prince Vasiliy III. Since then, Russian Tsars pretended to be Crown Princes of the Byzantine Empire, and Moscow was declared *the Third Rome*. As the first step in their attempt to reestablish the Christian Byzantine Empire, Russian Tsars tried to gain free access to the warm Black Sea. Catherine the Great was lucky enough to accomplish this first stage at the end of the 18th century. The huge newly conquered area along the northern coast of the Black Sea from the Danube River's delta (Izmail) to the Don River's delta (Rostov and Taganrog) was called Novorossia (New Russia). A dozen new ports and cities (Odessa, Kherson, Sevastopol, Novorossisk, Ekaterinodar, Rostov, Taganrog, etc.) were established. The Crimean Peninsula also was conquered, and thousands of families were brought from different Russian regions to settle there forever. The Empress (under Voltaire's influence?) granted to Novorossia something absolutely new for Russian history: Freedom. Serfdom was rejected here; moreover, any serf brought into this region by his lord was automatically declared a free man. Here, for the first time in Russian history, the Jews could become landowners without conversion to Christianity. Slavery for the Crimean Khans was declared outlaw in Novorossia, but the number of mosques, their shape, and elevation weren't limited as they were in the rest of Russia. And here, the Empress granted the Cossacks autonomy.

In response to these good graces, over the next 10 years, this huge and richly endowed (in natural resources) region began to prosper; and, unlike the rest of Russia, this prosperity continued until the Civil War (1918–1922); and even later—after the Cossacks were annihilated by the Communists following World War II, after Khrushchev's doubtful innovations in agriculture, after Brezhnev's incompetent leadership, after Gorbachev's perestroika and Yeltsin's idleness—this region has remained the most prosperous in the entire country.

In the middle of the last century, the eastern (Caucasian) coast of the Black Sea was conquered (also after heavy fighting with Turkey).

The Russian Empire acknowledged only two national entities—the Emperor entitled himself as Tsar of Poland and Grand Prince of Finland. Even such ancient states as Armenia and Georgia were not recognized as national entities inside the Empire. These two countries were divided into several provinces—as the entire Empire was—and the Tsar himself appointed the governors for each province.

The Communists, on the other hand, insisted on justice and complete equality of rights for all national minorities living in the Russian Empire. Therefore, Mr. Stalin changed the composition of the Empire by creating 16 (national) Soviet Republics, 8 of which constituted the Autonomous Republics or Autonomous Regions or Autonomous Districts (depending on the size of the population). And, the native populations were declared *socialistic nations*.

Thus, the eastern coast of the Black Sea fell to Georgia and its Autonomous Republic of Abkhazia, and the northwest corner of the Black Sea fell to the Ukraine.

In 1954, celebrating the 300th anniversary of the alliance of Cossacks under Getman[1] Khmelnitsky with Russian Tsar Alexey (official propaganda declared this event as the date when Ukraine voluntarily joined Russia), N. Khrushchev suddenly decided to give the Crimea to Ukraine. The anniversary above was false and artificial since: a) Ukraine was never a state; b) in 1658, the Cossacks, under a newly elected Getman Vigovsky and in an alliance with the Tatar's Khan of Crimea, started an extremely successful war against Russia and won it; c) Ukraine was finally joined to Russia under the rule of Empress Catherine the Great, at least 100 years later; d) this didn't happen of Ukraine's own accord, but it was a result of the partition of the Kingdom of Poland between Germany, Austria, and Russia.

Of course, the Crimea had never before belonged to Ukraine. Mr. Khrushchev's gesture might be excused if reviewed simply as a consequence of Stalin's mistake. For some clearly political reasons, Stalin gave the large western portion of the Cossacks' Steppes to Ukraine so that Russia lost its land contact with the Crimea. For N. Khrushchev, it seemed logical to give the Crimea to Ukraine in order to avoid various administrative troubles.

1. An elected Commander-in-Chief of the Cossack Army.

"He was a peaceful man, but he almost started two nuclear wars: the first with America, when he wanted to defend Cuba; the second with Ukraine, which wanted to defend the Crimea. Will we avoid this the second time? His grandchildren pray about it."

(M. Khrushcheva, N. Khrushchev's daughter-in-law; conversation with author, Moscow, June 1993)

"The Crimea can be either Ukrainian or inhabited."

(D. Korchinsky, Deputy Chairman of the Ukrainian National Assembly;[2] said at a meeting in Kiev, February 1994)

"The Crimea stays Russian, otherwise Ukraine will have to be annihilated."

(*Russkaya Rech*, daily newspaper; Petersburg, November 23, 1993)

◆

Almost 75% of the Black Sea coastline belonged to the FSU (Turkey, Bulgaria, and Romania own the rest). Once it was a resort zone along the seacoast stretching 2,500 miles and containing a dozen large ports with navy and submarine bases. Now, Russia owns less than 25% of the Black Sea coastline (Ukraine, Georgia, and Abkhazia own the other 50%) with only one large trade port (Novorossisk) housing an oil terminal. The other oil terminal is located in Tuapse and has almost four times less capacity. In the summer of 1993, a line of 36 tankers could be seen at the roadstead of Tuapse, each waiting for its turn to be loaded with Russian oil.

◆

"It takes us 2–3 days to feed a tanker. We could load 100 tankers per year but our schedule is for 280 tankers for this year. Everybody knows that we cannot keep to such a schedule, but ships keep coming. Are they coming just to stand for a while? Aren't they busy? . . . I can't help; Moscow can't help; God can't help."

(E. Tabakova, Chief of the Technical Department, Port Authority of Tuapse; conversation with author, Tuapse, June 1993)

2. An organization of the Ukrainian nationalists.

Meantime, in Tuapse and Novorossisk any prostitute could (for only $100) introduce you to the individuals who would be able to help. The mafia can help and even though their services are costly, it is still cheaper than the demurrage.

But for a few days each month, even the mafia was helpless because there was no oil in the terminal. Somewhere in Chechen, or in the north or in the south, the pipeline was tapped to siphon off as much oil as possible under the circumstances. Then, this oil comes to the port terminals in railway tanks, and the proper documents are presented. A tanker moves into position to be the first one fed, and the local mafia suddenly becomes very patient.

Oil had been stolen directly from pipelines all across Russia, but the greatest amounts were stolen within the Baltic Republics (Ukraine, Byelorus, Chechen, and Georgia).

There are a few reasons why the Russian government's reaction was and is so soft both toward the ports' mafia and pipelines' *oil recovery*. First of all, it is due to the constant bribery and corruption; then, the confusion in the ports and the muddle along the pipelines is of great help to the oil and transporting companies in trying to hide their semi-legal sales[3] of oil.

In 1993, the Russian police succeeded in uncovering almost 9,500 cases of illegal exports of raw materials and oil from Russia's ports. (This number is twice as much as it was in 1992.) These smugglers tried to export almost 18,000 metric tons of metals, 100,000 metric tons of oil, and 160,000 cubic meters of timber. The customs authorities say that it was less than 20% of the total contraband goods illegally transported from the country.

The smugglers' organizations have their agents within oil producing and transporting companies, mines' and plants' personnel, railway and port authorities, etc., and they have their *windows* at the state's boundary.

In 1993, among more than 8,000 people jailed for possession of contraband, almost 5,000 individuals were office employees (900 military and police officers).

> "Nobody wants or needs strict order today. It makes me sick to hear that this government is going to privatize the oil industry. It has almost been accomplished by the people themselves, and this government's duty is only to await a signal for the

3. This procedure may become the theme of a large, specialized book. If the government is truly serious about privatization, it should study this experience of self-privatization.

time when all this has to be approved—legally. They are smart if they can understand this. Then strict order will be reestablished, believe me. And it must happen rather soon. Otherwise, it will never happen."

(N. Uspenskaya, former Economics Advisor to Soviet Prime Minister J. Rizjkov; conversation with author, November 1993).

The Government's reaction was strong (and semi-legal) against smugglers only when it had political reasons to fight.

An experienced pilot, with 20 years' experience navigating within the Kaliningrad port (Baltic Sea), unexpectedly ran his ship aground here. A special team came to set this ship afloat, and several Moscow customs officers happened to be included in this team. They found that the ship was completely loaded with smuggled copper. The copper belonged to a Kazakhstan quasi-state firm.

A Cypriot tanker was loaded board-to-board (by several small tankers which followed the Volga River, Volga-Don Channel, and then entered the sea) in the middle of the Sea of Azov. Although the tanker was loaded with smuggled mazut[4] from Tatarstan, the Coast Guard kept silent because of a bribe. When this tanker appeared near Istanbul, it suspiciously caught on fire.

In late 1993, the Russian government failed when trying to set a priority for urgent oil shipments to the government's foreign clients. It was an alarming situation, so the government sent a signal back to the oil industry by postponing the issuing of the 1994 quotas until March. In January and February of 1994, the ports' oil terminals operated less hectically. And, the pipelines' *oil recovery* tactics almost stopped in Russia, also.

Had the mafia learned their lessons?

Rule 2–23

It should be clearly emphasized in each contract (agreement) that the amount and/or the price of oil you have bought or earned in Russia refers exclusively to the oil on-board the ship (FOB).

Let your Russian partners deal with the pipelines' owners and the ports' and terminals' authorities

4. Russian Mazut is roughly the equivalent of residual fuel oil, or resid.

◆

Ukraine's oil industry reached peak production in 1973, when 14.1 million tons of oil were produced. In 1991, only 4.9 million tons were produced, which is less than 10% of Ukraine's demands.

From 1968 to 1972, only the large gas field Shebelinka (in-place reserves were up to 625 billion cubic meters) located in the northeast corner of Ukraine, produced up to 30 billion cubic meters of natural gas annually. More than 75% of this gas was transported to Russia. In 1992, the total gas production of Ukraine was 100 times less.

Ukraine's major oil and gas fields in the Pre-Carpathian Foredeep and Dnieper-Donetsk Depression are located inland. The Crimea's oil and gas production input is rather modest. It produces almost 5% of the Ukrainian oil and less than 15% of Ukrainian gas.

Now in Ukraine, the political and economic situation is far worse than in Russia. And the conditions for foreign investments are really poor.

◆

> "It is now impossible to do business in Ukraine legally and make a profit. This is the situation now, but it could be better in three months."
>
> (W. Kish, Manager, Seagram; interview with a U.S. newspaper reporter, Kiev, January 27, 1994)

The situation did not get better.

In 1992, the Ukrainian Parliament granted tax holidays and other exemptions to draw foreigners. The original law stipulated that these conditions would not change for five years. Only six months later, however, many of the terms were scrapped. Then, in late 1993, the government suddenly suspended the weekly currency auctions and set a fixed rate on money conversion.

> "Believe me, I envy my Russian colleagues since they have their salary in rubles. You know that I never was a nationalist even though I thought that Ukraine would feel better being independent and free. Now, for me, the free Ukraine means: free from *this* government. Otherwise, we'll just cease to exist. . . . Yes, we need turboexpanders, gas plants, pipes, rigs, exploration, and everything else, but I would

not invite somebody to work with us. A foreigner will choke with anger dealing with our bedraggled authorities."

(V. Oleksuk, Chief Geologist, *Ukrgasprom* [Ukraine's Gas Industry]; conversation with author, Moscow, December 1993)

◆

Some large and mid-sized American oil companies showed interest in oil and gas fields of the Pre-Caucasian Basin, which dives under the Black and Azov Seas' water in its northeast corner.

The oil and gas fields of the Krasnodar (Ekaterinodar)[5] region are depleted. But those which are deep (4.5–6.0 km) and with low permeability of reservoir rocks still contain a lot of gas, condensate, and oil. Their development, by use of secondary recovery methods, might be efficient due to their location close to the sea ports and the excellent infrastructure within this region.

The two examples below on the scope of operations to be performed and the expected results might demonstrate the potential of secondary recovery in this region.

◆

Severskoe Field

- Well workover: 8
- Well drilling: 12
- Increase of daily production of gas: up to 6 million cubic meters
- Increase of daily production of condensate: 1,780 tons

East Severskoe Field

- Well workover: 3
- Well drilling: 6
- Increase of daily production of gas: 2 million cubic meters
- Increase of daily production of condensate: 400 tons

I'd give a high estimation to the oil reserves within the unexplored highland in the Krasnodar region. We have seen that in the eastern edge of the same basin (in Chechen), a few large oil fields were discovered within the highland.

---

5. *Ekaterinodar* (the former name) means in Russian: the gift from Catherine; *Krasnodar* means: the gift from reds (from the Bolsheviks).

The Black Sea and Azov Sea offshore areas are promising also.

In the Black Sea, the eastern offshore zone which belongs mainly to Russia is located between the coastline and the depth contour of approximately - 100 m, and it is relatively narrow (varies from 1.5 to 25 kilometers, Fig. 9-1). The reasons for positive estimation of the hydrocarbon potential of this zone are the following:

- The zone is located between two petroleum provinces (Crimea on the west and the Caucasian Foredeep on the east) with very similar geological features.
- The absence of very large fields within both provinces might be explained by absence of large anticlinal structures there.
- Almost 20 structures were mapped by seismic survey offshore, and 5 of them are larger than 15 km x 5 km (Yakornoe, Vernadskogo, Skalistaya, Mayachnaya, and Pionerskaya). These might be subject to priority attention.
- The only well drilled inside the zone brought salt water from the depth interval of 1,290 meters.
- There is not one well drilled deeper than 1,800 meters.
- All the structures inside this narrow zone are connected with the down-thrown blocks (steps) of the North Chernomorsky Rift (graven).

The geological conditions are very propitious for oil (gas) accumulation.

The western offshore zone belongs to Ukraine and, partly, to Romania. It is located in the very northwest corner of the Black Sea (Fig. 9-1) and is as large as almost 20,000 square kilometers. In 1988, oil fields were discovered here, and Romania started development in 1992.

During the last 7 years, 5 gas fields were discovered in the Ukrainian sector (Shtormovoe, Odessa, Krymskoe, Golitsino, and Shmidt); the rate of condensate varied from 25% to 45%.

More than 30 structures were mapped here by seismic survey but only 6 were drilled. The status of the other structures is unknown. At least 17 structures are as large as 55 square kilometers. The deepest hole reached the depth of 2,396 meters, while other wells aren't deeper than 1,860 meters. Not one hole uncovered the Lower Neogenic (N1) intervals.

A few gas fields were discovered offshore in the Azov Sea. A seismic line (Fig. 9-2) across the Oktyabriskoe gas field shows bar

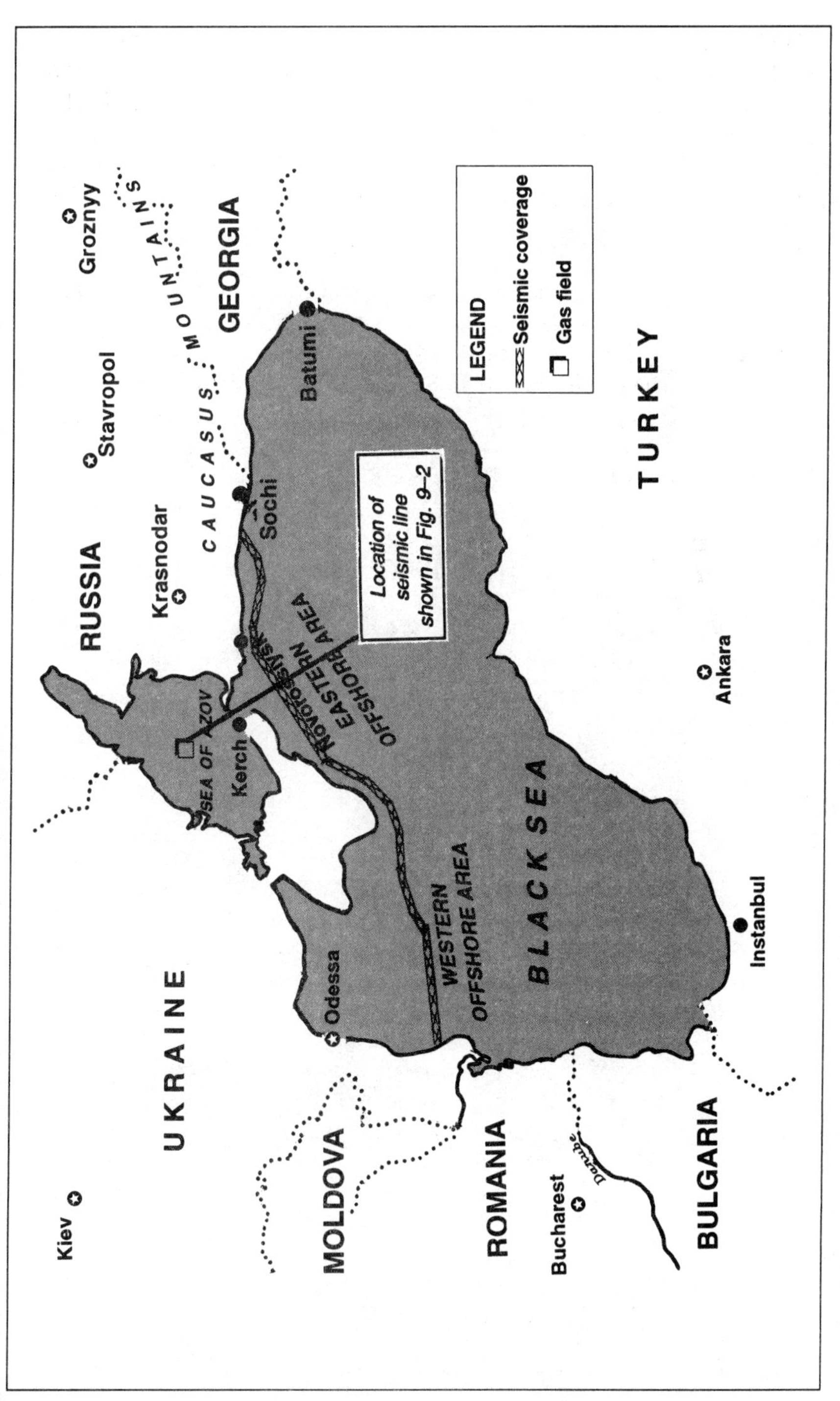

Figure 9–1 Eastern Russia's and Western Ukraine's Offshore Areas in the Black Sea

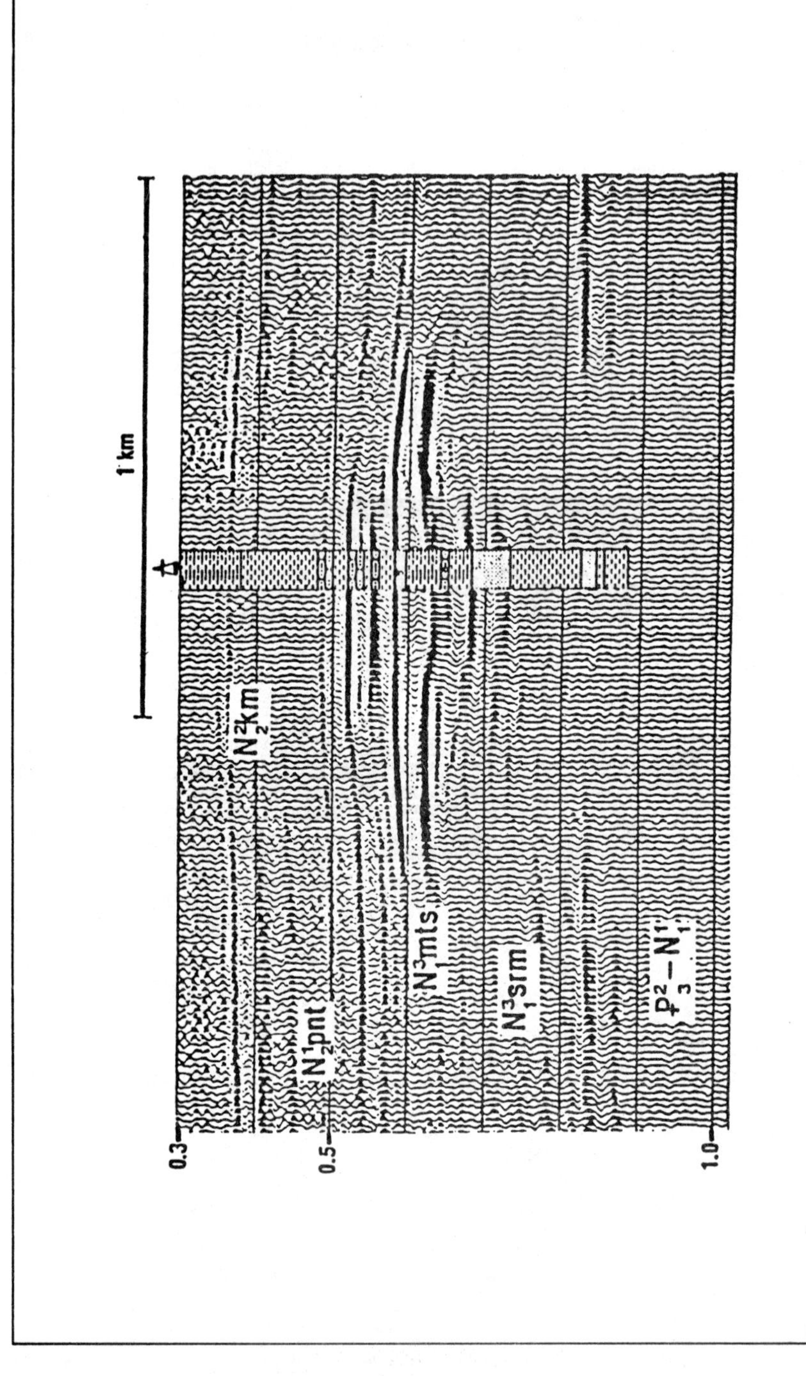

Figure 9–2 Seismic Line Across the Octyabriskoe Gas Field in the Sea of Azov

sandstones within a distal (shallow water shelf) sedimentary system of the Meotian time (N1mts). The sands and sandstones occurring in the intervals and represented on the seismic section by the most strong reflections contain gas and condensate (25%–35%).

> "I believe that the first deep hole drilled offshore in the Black Sea would discover a large oil field. I need one well as deep as 5 kilometers to turn the Crimea into a second Kuwait. Find an American as fanatical and we would have done it together. . . . Then don't worry, then we will have all the Black Sea navy defending us. This is Russia."[6]
>
> (V.N. Naydenov, Chief Geologist, *Chernomorneftegas* [Black Sea Oil and Gas]; conversation with author, Moscow, March 1993)
>
> "I'd suggest a point for a deep hole. I am almost sure that there is oil there. Neither Russia nor Ukraine have the money, technology, and equipment; nor will they have these even if they work together. And forget politics. Have you seen those thousands of Ukrainian peasants which sell food around Moscow's railway stations? They would rather feed Moscovites than their rulers. Look, these peasants vote each night when they take trains to Moscow to bring the food here. And they vote against nationalism. They will win. And we will win if we find oil offshore in the Black Sea."
>
> (V.Kuznetsov, General Director, *Yuzjmorneftegeofizika* [Southern Marine Geophysics]; conversation with author, Moscow, March 1993)

6. He lives in the Crimea. More than 60% of the peninsula inhabitants reject the idea that Crimea belongs to the Ukraine.

The Russian-oriented parties won the elections held in Crimea in early 1994. The newly elected President J. Meshkov visited Moscow immediately and brought back with him to Crimea a Russian citizen to be appointed as the Crimean Prime Minister. The most momentous events are still ahead.

> "There is no doubt that sometime in the future there will be occasions when the United States will have to look at Ukraine and look at Russia and say, 'Who do we choose?'"
>
> (R. M. Nixon, former President of the United States; said in a news conference, Kiev, March 16, 1993)

Well, either Russia or Ukraine has to be chosen sometime, and the choice will, in a certain proportion, depend on whether American business stays in Russia or in Ukraine. But American businessmen have to choose the correct strategy to stay away from the political games which are to be played by Russia, Ukraine, and the United States.

Chapter 10

This Is Russia: Rules, Comments, Advice

Some of these rules, comments, and advice can be applied to the Moslem countries of the FSU as well as Russia. The situation is the same for the rules given in Part 1, also. When reading both parts (Moslem and Russian), you can easily mark the specific (regional) rules and those which are more universal.

As in Part 1, these rules are not given in numerical order but by type and order in which they would be used by a company establishing business relationships in the FSU.

Rules

Reputation

Don't seek to solve your domestic financial problems in Russia.

Working in Russia, you will need solid reserves of time and money to eventually make a profit. (Rule 2-22, Chapter 8)

When working with Russians, use the same moral and ethical standards you use with your fellow Americans.

This isn't an issue of spiritual values or reputation only. In the Western world, the law helps to defend your interests, and, probably, you could work with somebody even though you don't trust him. Russians don't have this luxury. But they have the possibility to choose partners since competition is rapidly rising. (Rule 2-5, Chapter 6)

From the very beginning, keep the high reputation of your words. It will pay back much since it means a lot in Russia where this issue is a great rarity.
Russia isn't closed to dishonest people, entities, or intentions at all, but on average, the chances for them to succeed aren't much higher than in America.

Choosing Partners

Although the representatives of each of these three groups may become your partners, the local operating oil companies are in preference to stay with.

Being immediate producers, these local companies badly need all kinds of equipment and machinery, and they would try to avoid any middleman—Russian or American. Their needs have to be understood (if not respected) even in Moscow. (Rule 2-8, Chapter 6)

Contact state (quasi-state) enterprises since the logistics and financial rules regarding their relations with foreign investors are still much more clear and reliable.

Quasi-state means all kinds of joint stock companies, associations, etc., which were transformed from the state's enterprises by changing the name—nothing more important has been changed in their status; they remain state-owned. (Rule 2-2, Chapter 6)

Contact those private enterprises which have been formed by the heads of state-owned enterprises for better opportunities for financial maneuvers and/or for personal business reasons.

Almost each state-owned enterprise has such *private* partners. Obviously, the friends, relatives, or people with proper influence manage these kinds of enterprises. (Rule 2-3, Chapter 6)

The personal interest is a powerful factor in Moscow and the Russian provinces.

Stay away from those newly established private enterprises which aren't specialized yet and are looking for a niche for their activities.

It is almost impossible to determine what kind of *dirty* money they own if they do own any. Their contacts in the government might be based on bribery; plenty of agents representing gangs of racketeers serve therein. (Rule 2-4, Chapter 6)

Checking Your Partners

Try to check the origin of the money your Russian partners own, and involve them in joint operations.

Since no law system exists there, it is not only a warning against using *dirty* money. The problem is that this money may be a cover for another manipulation of the truth of which you'll never know. Or it will be too late for you to know. (Rule 2-12, Chapter 6)

If you are interested in a field, check the reserve figures; check the initial data and methods used to estimate the reserves. Check the number of wells drilled; check the number of productive wells; test the production rate.

Usually the shown materials are wrong and nobody remembers when and why the data were changed. Since the drillers' salary depended on the weekly footage, even the depths report might be wrong. (Rule 2-7, Chapter 6)

Working With Your Partners

Hire a Russian lawyer to prepare all the documents needed to register your joint venture.

Do it by yourself (hiring the lawyer, preparing the documents, and registering the joint venture). You can be sure that your Russian partners will not have done this job properly. Also, they might try to bribe clerks secretly. (Rule 2-21, Chapter 8)

Don't hunt for fields located in the far corners (*bear's corners*).

You cannot predict the problems to be faced in this type of venture. (Rule 2-19, Chapter 8)

All the data and information you may need is available in Moscow.

Check the figures which your Russian partners report in the documents.

They will try to exaggerate your achievements. Advise them to be modest. (Rule 2-20, Chapter 8)

A Strategy of Unfranchised Status

When doing business, follow the laws and regulations which have been chosen to be observed by your Russian partner.

This isn't a choice between some *good* or *bad* laws (or decrees); this is a choice of people who control the observance. Your partner knows these people better. (Rule 2-6, Chapter 6)

Avoid getting involved in political fights, but stay ready, together with your Russian partners and allies, to defend your lawful interests.

In modern Russia, at least an attempt has become possible. (Rule 2-9, Chapter 6)

If your American experience contradicts your Russian adviser's suggestion, follow the latter.

Most often you are trying to follow *your* common sense; Russians have their own since different factors determine it. You may know these factors but misunderstand their importance. (Rule 2-13, Chapter 6)

Don't appeal to the high political authorities with your business problems. Look for a proper key.

In modern Russia, it is almost impossible to tell who is who in different businesses (especially in the financial mafia)—even for the highest political leaders. So the latter would be very careful in entering all kind of business squabbles. The *proper key* means proper people and their interests. (Rule 2-10, Chapter 6)

Money

Use proper financial terms when discussing business. Avoid using the word money. Use the word cash when you have this in your mind.

This is the way to avoid a lot of confusion and misunderstanding. (Rule 2-1, Chapter 6)

Don't expect a fast money turnover.

Even some trade companies got stuck on this issue, which became a real national disaster. It is difficult to improve this practice since several very powerful groups make their profit by the turning over of money which doesn't belong to them. (Rule 2-11, Chapter 6)

It should be clearly emphasized in each contract (agreement) that the amount and/or the price of oil you have bought or earned in Russia refers exclusively to the oil on-board the ship (FOB).

Let your Russian partners deal with the pipelines' owners and the ports' and terminals' authorities. (Rule 2-23, Chapter 9)

Security

Avoid conducting any financial operations in cash, either rubles or currency.

Criminal groups try to have special informers everywhere, including different banks. (Rule 2-18, Chapter 7)

Be sure your shipments are well guarded when crossing Russia by rails or trucks. Be sure the storage of your equipment and your office(s) are well guarded. It pays to carry these large expenses.

Avoid using the police either as security guards or as detectives. Instead, try to enlist the services of private companies to provide these functions.

The criminal groups try to have special informers in the police. Those private companies, however, have their own mechanism of keeping the gangs at a distance by intimidation and clear consequences of violating the code. (Rule 2-17, Chapter 7)

Be sure your guards belong to a legally established security firm.

Alcohol

Russians like to drink, and then they like to keep filling in details about the drinking. This is their way to cure stresses, also.

At the table, never refuse to drink.

This is a matter of personal relationship. You better empty your glass into your shoes or leave the party in the middle. Your absence at the table from the very beginning would be pardonable also. (Rule 2-16, Chapter 7)

Don't bring with you persons with alcohol-related problems.
Russia is not the place for them to work. (Rule 2-15, Chapter 7)

Screen each potential employee for alcohol-related problems.
Your direct question could be enough since Russians don't consider this addiction as a vice. But a Russian will never admit to being a chronic alcoholic, especially if it's true. Also, make sure that the potential employee is an excellent worker during periods of boozing it up. (Rule 2-14, Chapter 7)

Index

A

Abkhazia 73, 75, 178, 179
Black Sea 179
independence from Georgia 75
accommodations 113, 114-115
communications 114
entertainment 114-115
exchanging currency 115
hotels 113
meals 114
shopping 115
advertising 99
information about your company (Rule 1-38) 99, 106
agents and partners for Western businesses 27, 28, 33, 41, 64, 67-69, 79, 102, 108-109, 110, 112, 113, 115, 129, 130, 134, 135, 140, 150, 151, 153, 167, 170, 181, 189, 190, 191, 193
choosing partners 190
dealing with port and terminal authorities 181
Rule 1-3 27, 110
Rule 1-4 28, 108
Rule 1-8 33, 111
Rule 1-15 108
Rule 1-16 108
Rule 1-17 41, 112
Rule 1-20 109
Rule 1-25 64, 112
Rule 1-27 67, 112
Rule 1-31 79, 111
Rule 1-32 79, 111
Rule 1-33 109
Rule 1-40 102, 113
Rule 2-2 126, 190
Rule 2-3 127, 190
Rule 2-4 127, 190
Rule 2-5129, 189
Rule 2-6 130
Rule 2-7 191
Rule 2-8 134, 190
Rule 2-9 135
Rule 2-12 140, 191
Rule 2-13 140
Rule 2-14 150
Rule 2-15 151
Rule 2-16 151
Rule 2-17 153, 193
Rule 2-18 153, 193
Rule 2-19 191
Rule 2-20 167, 191
Rule 2-21 170, 191
Rule 2-23 181, 193
Akaev, Askar (President of Kirgizstan) 13, 14, 19, 28, 29, 32, 43-44, 45, 46, 87
alcohol-related problems 150-151, 193-194
crime 151
Rule 2-14 150, 194
Rule 2-15 151, 194
Rule 2-16 151, 193
alphabets 48
Altay Mountains 3
aluminum 14
Amoco 120, 135, 174
Harvey, J. (DLJ Securities analyst) 174
Morland, A. (Vice-President) 120
Timan-Pechora 135
Amu-Darya River 47
Andropov, Yuri (KGB head, General Secretary of FSU's Communist Party) 52
Angola 125
Antarctica 45
antimony 14, 125
application for joint venture (Memorandum of Association) 169, 191
Rule 2-21 169, 191
Aral Sea 7, 52
pollution 52-53
Arctic Ocean 157
Armenia 18, 47, 73, 73, 178
Communism 47
ASARCO 17
Astrakhan 134-135, 139

oil deal 134-135
Ataberdiev, M. (Turkmenistan's Deputy Minister of Fuel) 67
Austria 167, 178
Autonomous Republic of Khanty and Mancy 160
avoiding troubles 110-111, 166, 167, 170, 171, 189, 191
Rule 1-3 110
Rule 1-12 110
Rule 1-13 110
Rule 1-31 111
Rule 1-36 110
Rule 2-19 166, 191
Rule 2-20 167, 191
Rule 2-21 170, 191
Rule 2-22 171, 189
Aytmatov, Chingiz (Kirgiz writer) 23
Azerbaijan 78, 140, 141-142
nationalism 142
offshore oil fields 141-142
Apache oil deal 141-142
British Petroleum (BP) 141
Major, John 141
Thatcher, Margaret 141
presidents 140
Aliev, Geydar (KGB Chief and First Secretary of Communist Party) 141
Elchibey 140-141
Mutalibov 140, 141-142
South Caspian Basin fields 81
Caspian Sea offshore 82
Onshore 81
Azov, Sea of 181, 183, 184, 185, 187
oil and gas fields 183, 184-187
Octyabriskoe 184-187
map of 186

B

Babilashvily, A. (University of Tbilisi Professor of History) 76
Baker, James (U.S. Secretary of State) 7, 8
Balkans 76
Bally (Rice University Professor) 82
seismic surveys 82
Baltic republics 20, 94, 141
history 94
nationalism 141
population 94
Baltic Sea 181
Kaliningrad, port of 181
Barents Sea offshore oil field 82, 87
Bashkiyra 159
Baybakov (Oil Minister; Prime Minister; Chairman of *Gosplan*, State Planning Committee) 79
behavior, safe 67-68, 79, 112, 113
Rule 1-22 112
Rule 1-28 68, 112
Rule 1-30 79, 112
Belyakov, V. (Senior Geologist, *Samotlorneft*) 161
Belov, V. (President of Russian refinery) 127-128
Alexander (employee) 127-128
Protcenko (Vice-President, Chief of Moscow Regional Press Club) 127-128
failed deal 127-128
Beria 73
freed criminals 73
Birshtine, B. (associate of Kirgizstan President Akaev) 32
bitumen, solid 95-96, 104
black markets, avoiding 115
Black Sea 20, 43, 74, 76, 84, 141, 167, 169, 177-188
bordering nations 179
Chernomoroneftgas (Black Sea Oil and Gas) 187
Naydenov, V. N. (Chief Geologist) 187
oil 152, 179, 180, 181, 183, 184
map of offshore areas 185
oil and gas fields 183, 184
ports of 43, 74, 84, 141, 167, 169, 179, 180
theft of oil 180-181
blowout preventers (BOPs) 49
BMB (Turkish gold company) 55
Bokhara 47, 49, 113
accommodations 113
Bolsheviks 183
Bratsk 163
Brazauskas, A. (President of Lithuania) 28
Brezhnev, Leonid 52-53, 79

Gorbachev, M. (son-in-law, former Deputy Minister of Interior Affairs) 52-53
Baybakov (Prime Minister) 79
bribery 38-40, 101, 102, 113, 115, 180-181
Rule 1-40 102, 113
Russian Coast Guard 181
Britain 7, 12, 13, 56
Scotland 56
National Westminster Bank of Scotland (NatWest) 56
British Gas 154
Karachaganack oil field 154
British Petroleum (BP) 141-142
failed oil deal 141-142
Thatcher, Margaret (representative) 141
Brown, R. H. (U.S. Secretary of Commerce) 178
Brush Creek Mining and Development Co. 17
Buckley, P. (Vice-President of Petron, Inc.) 29, 81
Burburlis (B. Yeltsin's associate) 128
Bush, George 43, 122
business starting 98, 110, 161-162, 170-174, 189, 191
Rule 1-36 98, 110
Rule 2-21 170, 191
Rule 2-22 171, 189
Byelorussia 67, 82, 94, 143
American Federal Reserve System to the Byelorus Republic 143
Laws, R. (Representative) 143
history 94
Kuznetsov (First Deputy Chairman) 143
oil industry 143
population 94
Pripyat Trough oil fields 82
Byelorusneftegeofizika 149
India contracts 149

C

cadmium 57
Cambre, M. (Newmont CEO) 55
Cameco 44, 45, 46
Khomenuk, L. (President of Kum-Tor Operations Company, Cameco's branch) 44, 46
Cameco (TSE) 17
Canada 7, 12, 28, 49, 55, 56
Caspian Sea 3, 20, 64, 66
oil fields on eastern coast 66
Caucasian mountains 71
Caucasians 152
Caucasus 3
charter of joint venture 170, 171, 191
Rule 2-21 170, 191
Chechen-Ingush Autonomous Republic 72, 78
Chechen Republic, the 71-91, 96, 177, 180, 183
agriculture 71
Armenia 73
Baybakov (Oil Minister; Prime Minister; Chairman of State Planning Committee) 79
borders 75
Chechen *mafia* 84, 91
Chevron 87
Christianity 71
Communism 72, 75
complaints about American oil equipment 91
Conoco 87
Cossacks 71, 73
annihilation of 73
Gorbachev 73
KGB 73
crime 73, 75, 87-88, 90, 113
Beria 73
freed criminals 73
Chechen *mafia* 84, 91
VIP 91
Communism 75
Shevardnadze, E. (FSU Foreign Minister, Georgia's KGB Chief, First Secretary of Communist Party) 75
coup d'etat in Georgia 75
theft of oil money 90
culture 72
Cyprus 87
Deputy Prime Minister 84-86
Doudaev, Johar (President, former Soviet Air Force General) 75, 77, 80, 87-88

nuclear bombs 77
earthquakes 113
economy 78
impact of education 78
education 77-81
Baybakov (Oil Minister, Prime Minister, Chairman of State Planning Committee) 79
Equal Opportunities Policy 77-81
impact on economy 78
Institute of Combustible Fossils 79
results of the Russian program 77-81
Finance Minister 88
Galburaev, Ruslan (Chief Engineer of Sakhalin Oil, Minister of Chechen Republic's Oil Industry) 74, 80, 83, 84-86, 91
Gamsakhurdia (President) 72, 73, 74, 75, 76, 80, 86, 88, 91
German army 72
Gorbachev, M. 73, 75, 87
Gosplan (Russian State Planning Committee) 79
Baybakov (Chairman) 79
Grozneftegeofizika (local oil operating company) 82, 149
Peru contracts 149
Grozny (capital) 72, 73, 74, 75, 80, 81, 86, 88, 91, 113
apartments 113
crime 73, 113
dachas 113
earthquakes 113
Eastern Region oil fields 81
German army 72
hotels 113
Nobel's oil fields 86
Grozny Petroleum Institute 80
history 71-72, 73, 75, 76-77
hospitality 71-72
independence 72, 75, 76
Ingush 72
IRA (U.S. oil company) 85
Islam 71, 74
Ivan the Terrible 77
Jews 73
KGB 73, 75
Khasbulatov, Rouslan (Speaker of Russian Parliament) 88-89
Krasnodar 81
Western Region oil fields 81
Khrushchev, Nikita 72, 79
Baybakov (Oil Minister) 79
language (Germanic) 72
oil industry
complaints about American equipment 91
Cretaceous deposit 83
exploration 81, 87
seismic surveys 82
with submarines 87
Jurassic deposit 83
Lower Neogenic sediments 83
negotiating oil contracts 84-86, 88-89
oil fields 81, 82, 86, 87, 149, 183
Barents Sea offshore fields 82, 87
Mangyshlack Basin 82
North Caucasian Region 81
Central Region 81
Eastern Region 81, 86
Nobel, Alfred 86
Western Region 81
Peru contracts 149
Pre-Caspian Basin 20, 82, 153-154, 183
Pripyat Trough 82
Sakhalin Basin 82, 87
contracts 87
South Caspian Basin 81
Caspian Sea offshore 82
Onshore 81
Volga-Ural Basin 82
West Siberian Basin 82
Oil Ministry in Moscow 96
Paleogenic sediments 83
Paleozoic reef 83
Triassic 83
pipelines 74, 180
refinery 73
Russian specialists 86
seismic surveys 82
status of industry 80
theft of oil and money 90, 180
Upper Cretaceous limestones 83
World War II 72
Orujev (FSU Deputy Minister of Oil) 91

political conditions 73-76
Sokolovsky, Professor E. V. (Director of the North Caucasian Petroleum Research Institute) 73-74, 84-86
United States involvement 75-76
Ames, (CIA Officer and Russian Spy) 75
population 72-73
Sakhlainneft (Sakhalin Oil) 80, 83, 84-86, 91
Shevardnadze, E. (FSU Foreign Minister, Georgia's KGB Chief, First Secretary of Communist Party) 75, 84
coup d'etat in Georgia 75
Siberia 72, 74
slavery 72
Stalin, J. 72, 73, 76
Stavropol 74, 81
Central Region oil fields 81
traditional occupations 71
Urazjtcev (member of Russian Parliament) 88-89
vendettas 71
Vietnam 74
Yeltsin, B. 73, 89
Yermolov (Russian General) 72
Zhirinovsky 74
Chengyshev (Prime Minister of Kirgizstan) 40-43, 87
Chepursky, V. (Chief Engineer of West Siberian Oil Pipelines) 169-170
Chernomoroneftgas (Black Sea Oil and Gas) 187
Chernomydin, Z. (Russian Prime Minister, Minister of Gas Industry) 87, 133, 134, 135, 158, 175
Chevron 20, 43, 87, 148
contract with (Saratov Oil Geophysical Survey) 148
Tengiz oil contract 87
China 9, 12, 13, 18, 19, 21
Christopher, Warren (U.S. Secretary of State) 6, 7
CIA 75-76
Ames (CIA Officer and Russian Spy) 75-76
Clinton, William 122
coal 125
Russian resources 125
Commonwealth of Independent States (CIS) 124
communications, availability of 114
Communism 12, 16, 20, 28, 30, 47, 52, 57, 72, 73, 75, 102, 123-125, 133, 134, 135, 136, 138, 177, 178
Andropov, Yuri (KGB head, General Secretary of FSU's Communist Party) 52
Armenia 47
concept of money 123-125
crime 75
Beria 73
freed criminals 73
Shevardnadze, E. (FSU Foreign Minister, Georgia's KGB Chief, First Secretary of Communist Party) 75, 84
coup d'etat in Georgia 75
equality of citizens 178
Gorbachev, M. (General Secretary of Communist Party) 136
Marx, K. 123
Moscow Committee of the Communist Party 136
Political Bureau of the Central Committee 136
Rashidov, S. (First Secretary of the Central Committee of Uzbekistan's Communist Party) 52-53
religious beliefs 102
von Bismarck, O. (Chancellor of the German Empire) 123
Ziuganov, G. (Chairman of the Russian Party) 133, 161
Yeltsin, B. (Russian President) 136
ideas about democracy 136
Competition in Incompetence 7
conducting business 23, 26, 31, 33, 38, 41, 42, 44, 61, 64, 166, 167, 170, 171, 181, 189, 191, 193
conducting business meetings 38, 41
Rule 1-18 41, 107
Rule 1-13 38, 110
introductions, importance of 98

Rule 1-21 44, 111
Rule 1-40 102, 113
Rule 2-22 171, 189
personal relationships 23, 26
Rule 1-1 23, 106
Rule 1-2 26, 106 Rule 1-6 31, 107
Rule 1-8 33, 111
Rule 1-12 38, 110
Rule 1-14 41, 105-106
Rule 1-20 42, 109
Rule 1-23 61, 106
Rule 1-25 64, 112
Rule 1-36 98, 110
Rule 1-38 99, 106
Rule 1-39 101, 106
Rule 2-19 166, 191
Rule 2-20 167, 191
Rule 2-21 170, 191
Rule 2-23 181, 193
Conoco 87
Barents Sea offshore field 82, 87
contracts 38-40, 42, 44, 61, 64, 84-86, 87, 88-89, 101, 102, 113, 150, 151, 153, 167, 170, 180-181, 191, 193-194
bribery 38-40, 101, 102, 113, 115, 180-181
protocols 101
Rule 1-12 38, 110
Rule 1-13 38, 110
Rule 1-19 42, 107
Rule 1-20 42, 109
Rule 1-21 44, 111
Rule 1-23 61, 106
Rule 1-25 64, 112
Rule 1-26 64, 108
Rule 1-33 87, 109
Rule 1-34 87, 111
Rule 1-39 101, 106
Rule 1-40 102, 113
Rule 2-14 150, 194
Rule 2-15 151, 194
Rule 2-16 151, 193
Rule 2-17 153, 193
Rule 2-18 153, 193
Rule 2-20 167, 191
Rule 2-21 170, 191
Rule 2-23 181, 193
Control Risk Group (England) 121
copper 14, 57, 181
Cossacks 71, 73, 151, 152, 177, 178
annihilation of 73, 177
Getman Khmelnitsky 178
Getman Vigovsky 178
Gorbachev 73
KGB 73
Pugachev, E. (leader) 151, 152
revolt against Ekatherine II 151, 152
Tatar Khan 178
Couchougoura (Chief Engineer, Stavropol Geophysical Survey) 3
coup d'etat against Gorbachev (1991) 103-140
courtesy during visits to the FSU 115
Crimea 50, 93-94, 177-179, 182, 184, 187, 188
Khans 177
Kuznetsov, V. (Southern Marine Geophysics General Director) 187
Meshkov, J. (President) 188
nationalism 187-188
oil and gas production 182, 184
population 94
crime 38, 40, 64, 68-69, 73, 75, 87-88, 90, 102, 112, 113, 138-140, 151, 152-153, 167, 180-181, 193
alcohol-related problems 151
Beria 73
freed criminals 73
bribery 38-40, 102, 113, 180-181
Rule 1-40 102, 113
Russian Coast Guard 181
Chechen *mafia* 84, 91
Communism 75
fraud in ecological cleanups 167
Interior Affairs Ministry 153
in the Chechen Republic 87-88
theft of oil money 90
in Tashkent (capital of Uzbekistan) 68-69
mafia 84, 91, 153
Moscow 138-140
private security firms 152-153
Rule 2-17 153, 193
Rule 2-18 153, 193
prostitution 67-68, 112, 113, 180

Rule 1-28 68, 112
Saratov 151
theft of resources 18-181, 193
Rule 2-23 181, 193
crude oil 49
Cuba 179
Cyprus 87
Czechs 12

D

dachas 113
Dagestan 86, 87-88, 89
Dagneft (Dagestan Oil) 89
Jabrailov (Chief Engineer) 89
Sayidov (Chief) 89-90
education 81
financial dealings 87-88
Lisovsky (head of the Geological Management of the FSU Oil Minister) 89-90
seismic surveys 86
Danube 177
Darius (Shah) 47
Dautov, Marsel A. (Minister of Tatarstan Community Services) 99-103
de Larosiere, J. (EBRD President) 55
democracy 45, 49, 136
B. Yeltsin's ideas about 136
Denau train station 56
diamonds 16
discussing international relations (Rule 1-29) 71, 109
discussing political leaders (Rule 1-7) 32, 110
discussing politics 30, 109-110
Rule 1-5 30
Rule 1-7 110
Rule 1-29 109
discussing wives and daughters (Rule 1-11) 35
DLJ Securities 174
Harvey, J (Analyst) 174
Don 177
Doudaev, Johar (President of the Chechen Republic, Soviet Air Force General) 75, 77, 80, 87-88
nuclear bombs 77
dressing in the best style (Rule 1-14) 39, 105-106
Druzhba (Russian pipeline) 141, 169
Dunavant Cotton & Ginning Service 68
Zenina, Oksana M. (European Representative) 68
Dupre, M. (General Director, *Tengizchevroil*) 20
Dushanbe-Termez road (Uzbekistan) 56

E

earthquakes 113
safety precautions 113
EBRD 55
de Larosiere, J. (President) 55
economy 78
impact of education 78
education 71, 77-81
Chechen Republic, the 77-81
Equal Opportunity Policy 77-81
FSU 77-81
impact on economy 78
results of Russian program 71
Rule 1-32 79, 111
Egypt 142
oil fields 142
Ekaterinodar 177, 183
Ekatherine II (see *Catherine the Great)*
Elchibey (President of Azerbaijan)140-141
Elf Aquitaine 148
Saratovneftegeofizika contract 148
England 6, 87-88, 122
Control Risk Group 122
entertainment 114-115
Equal Opportunity Policy 77-81
etiquette 35, 105, 107
impact on business 35
Rule 1-6 107
Rule 1-9 107
Rule 1-10 107
Rule 1-11 35
Rule 1-18 107
Rule 1-19 107
Rule 1-35 96, 107
Rule 1-37 98, 105
Eurasia 9, 45
Europe 3

European Development Bank 55
Newmont 17, 29, 37, 39, 53, 55, 64
exchanging currency 115
excluding women from *Hoshars* (Rule 1-26) 64, 108
Export-Import Bank 161
expressing wealth (Rule 1-14) 39, 105-106
Exxon 135
Timan-Pechora 135

F

faxes, sending from the FSU 114
feasibility study of joint ventures 170, 171, 191
Rule 2-21 170, 191
Rule 2-22 171, 189
Federal Express 114
Fedynsky (Reporter for Voice of America) 12
Fergana intermountain basin 18, 19, 21, 23, 41, 43, 48-49
Kirgizstan 21, 24
oil drilling 50
oil-well explosion 62-63
refinery 49
Uzbekistan 53, 62
financial terms 123, 192
Rule 2-1 123, 192
Finland 93, 169, 178
First Uzbek Independent University of Business and Diplomacy 48
Tursunov, B. (Chancellor) 48
foreign investments 161-162, 170-174, 182-183
government programs 161-162
joint ventures 161, 170-171
laws concerning 172-174
Parliament (Soviet and Russian) 172-173
Ukraine 182-183
Yeltsin 173, 174
former Soviet Union (FSU)
Alexey, Tsar 178
Ames (CIA Officer and Russian Spy) 75-76
Andropov, Yuri (former KGB head and General Secretary of Communist Party) 52
anti-Americanism 176
Arkhangelsk region (northwest of European Russia) 120
army 93, 133, 144
Mokashov, A. (General) 144
World War I 93
World War II 93
Astrakhan region 134-135
oil deal 134-135
Baybakov (Oil Minister, Prime Minister, Chairman of State Planning Committee) 79
Belogorod region 120
Belov, V. (President of Russian refinery) 127-128
Alexander (employee) 127-128
Aliev, Geydar (KGB Chief, First Secretary of Azerbaijan's Communist Party) 141
failed deal 127-128
Protcenko (Vice-President of refinery, Chief of the Moscow Regional Press Club) 127-128
Beria 73
freed criminals 73
borders 75
Brezhnev, Leonid 52-53
Gorbachev, M. (son-in-law, former Deputy Minister of Interior Affairs) 52-53
Burburlis (B. Yeltsin's associate) 128
Catherine the Great (Ekatherine II) 151, 152, 177, 178
Pugachev, E. (Cossack leader) 151, 53
revolt against 151, 152
Suvorov, Generalissimus A. 151, 152
Chechen-Ingush Autonomous Republic 72, 78
Chechen Republic, the 71-91
dispute over independence 72, 75
Chernomyrdin, Z. (Russian Prime Minister, Minister of the Gas Industry) 87, 133, 134, 135, 158, 175
International Monetary Fund 175
Civil War 177

Coast Guard 181
Commonwealth of Independent States (CIS) 124
Communism 12, 16, 20, 28, 30, 47, 52, 57, 72, 75, 102, 123, 177, 178
  Andropov, Yuri (former KGB head and General Secretary of the Soviet Union's Communist Party) 52
  annihilation of Cossacks 73, 177 Armenia 47
  Autonomous Republic of Khanty and Mancy 160
  Chechen-Ingush Autonomous Republic 72
  concept of money 123
  crime 75, 138-140
    Beria 73
    Moscow 138-140
    Zhevradnadze, E. (FSU Foreign Minister, Georgia's KGB Chief, First Secretary of Communist Party) 75, 84
      coup d'etat inGeorgia 75
  equality of citizens 178
  Gorbachev, M. (General Secretary of Communist Party) 136
  Marx, K. 123
  Moscow Committee of the Communist Party 136
    Yeltsin, B. 136
  Political Bureau of the Central Committee 136
  religious beliefs 102
  Rashidov, S. (First Secretary of the Central Committee of Uzbekistan's Communist Party) 52-53
  von Bismarck, O. (Chancellor of the German Empire) 123
  Ziuganov, G. (Leader of the Russian Communist Party) 133, 161
coup d'etat against Gorbachev (1991) 103-140
crime 38, 40, 64, 73, 87-88, 102, 112, 113, 150-151, 153, 167, 180-181, 193-194
  alcohol-related problems 150-151, 193-194
    crime 151
    Rule 2-14 150, 194
    Rule 2-15 151, 194
    Rule 2-16 151, 193
  bribery 38-40, 102, 113, 180-181
    Rule 1-40 102, 113
    Russian Coast Guard 181
  fraud in ecological cleanups 167
  Interior Affairs Ministry 153
  mafia 84, 91, 153, 180
    theft of oil 180
  prostitution 67-68, 112, 113, 180
    Rule 1-28 68, 112
Dagestan 86, 87-88, 89
Dagneft (Dagestan Oil) 89
  Jabrailov (Chief Engineer) 89
  Sayidov (Chief) 89-90
democracy 136-138, 144
  Yeltsin, B. 136-138
Doudaev, Johar (President of the Chechen Republic; former Soviet Air Force General) 75, 77, 80, 87-88
economic and political situations 121-122, 139, 140
  Gallup poll 121
    reasons for difficult business situations 121
  inflation 139
    monetary policy 139
economy 78
  impact of education 78
education 77-81
  Equal Opportunity Policy 77-81
  impact on economy 78
  Institute of Combustible Fossils (IGIRGI) 70
  results of Russian program 77-81
Ekatherine II (see *Catherine the Great)*
Equal Opportunity Policy 77-81
Fergana intermountain basin 18, 19, 21, 23, 41, 43, 48-49
  Minbulak area 21, 24

foreign investments 161-162, 170-174, 182-183
government programs 161-162
joint ventures 161, 170-171
laws concerning 172-174
Parliament (Soviet and Russian) 172-173
Ukraine 182-183
Yeltsin 173, 174
forming a joint venture 170-171, 189, 191
documents to file 170
Rule 2-21 170, 191
Rule 2-22 171, 189
Gabrielyants, G. A. (Minister of Geology) 132
Gaidar, Y. (former acting Russian Prime Minister, President of Astrakhan oil company) 134, 139, 160
gold reserves (see *gold)*
Golovaty, A. (Deputy Minister of Finance) 173
Gorbachev, M. (General Secretary of the Communist Party) 48, 73, 75, 87, 103-104, 124, 133, 136, 147, 148-149, 177
coup d'etat against (1991) 103-140
privatization of businesses 147, 148-149
vouchers in FSU businesses 148-149
Tengiz oil contract with Chevron 87
Gorbachev, M. (former Deputy Minister of Interior Affairs, L. Brezhnev's son-in-law) 52-53
Gosplan (State Planning Committee) 79
Baybakov (Chairman) 79
government involvement in oil industry 96
Heroes of the Soviet Union 93-94
Herzen, A. I. (writer) 130
history 71, 72, 76, 93, 123-125, 144-145, 151, 152, 177-179
independent states formed from
Azerbaijan (see *Azerbaijan)*
Kazakhstan (see *Kazakhstan)*
Kirgizstan (see *Kirgizstan)*
Tajikistan (see *Tajikistan)*
Turkmenistan (see *Turkmenistan)*
Uzbekistan (see *Uzbekistan)*
Institute of Combustible Fossils (IGIRGI) 70
Intourist 30, 113
hotels 113
Jabrailov (Chief Engineer of Dagestan Oil) 89
Jakutiya (Siberian republic) 55
Jewish autonomous region (Russian Far East) 120
Kaganovich, L. (First Secretary in Kiev) 94
Karpov, A. (World Chess Champion) 127-129
KGB 28, 30, 48, 52, 73, 75, 133, 137, 152-153, 193
Andropov, Yuri (former KGB head, General Secretary of FSU's Communist Party) 52
private security services 152-153, 193
Rule 2-17 153, 193
Rule 2-18 153, 193
Khasbulatov, Rouslan (Speaker of the first Russian Parliament) 88-89
Komi Autonomous Republic 120, 135, 155
Texaco oil project 120, 135
Krasnoyarsk region 120
Khrushchev, M. (N. Khrushchev's daughter) 179
Khrushchev, Nikita 72, 79, 177, 178-179
nuclear war 179
Krylov, I. (Chief of Staff of the Representatives of the President of Russia) 152
Kuznetsov (First Deputy Chairman of the Byelorus Republic) 143
Lenin, V. I. 53, 77, 123
Lisovsky (former head of the Geological Management of the FSU Oil Ministery) 89-90
long-term issues for foreign investment 175-176
anti-Americanism 176

Lukoil (Russian company) 141-142
map of 4-5
mineral resources 125
Ministry of Ecology 167, 169
oil pollution 167
Ministry of Finance 173
Golovaty, A. (Deputy Minister) 173
Ministry of Fuel and Energy 133, 134, 168
Shafranik, Y. (Minister) 161
Taslitsky, E. (Deputy Minister) 168
Ministry of Geology 30, 131, 132, 134
Gabrielyants, G. A. (Minister) 132
Ministry of Oil 18, 26, 96, 131, 133, 134, 148
Central Geophysical Expedition 148
sale of information 148
Ministry of the Gas Industry 133, 134, 135
Chernomyrdin, Z. (Minister, Russian Prime Minister) 133, 134, 135, 158, 175
International Monetary Fund 175
Minsk 91, 143
Mokashov, A. (General of the Soviet Army) 144
money, concept of 123-125, 192
history 123-125
Rule 2-1 123, 192
Moscow (see *Moscow)*
nationalism 142-145
Nezavisimaya Gazeta (Moscow daily) 139
Novorussia (New Russia) 177
obtaining data 64
oil industry
American-Russian joint ventures 120
areas where western investment is encouraged 162-168
aspects of 122, 123-135
conditions of 121-122
exploration 87, 166, 175, 191
Rule 2-19 166, 191
with submarines 87, 175
exports 120
government and foreign investments 161-162
government involvement 96
Lukoil 141-142
Azerbaijan oil deal 141-142
Frankenburgh lawsuit 142
Mitcui loan 142
World Bank loan 142
world investments 142
oil fields 120, 134-135, 141-142, 184
Astrakhan 134-135
Azerbaijan 141-142
Apache oil deal 141-142
history 141
offshore oil fields 142
Azery 142
Gunashtly 142
Shirak 142
Komi 120
Texaco project 120
offshore zone 184
Roman Trebs 135
Rule 2-7 132, 191
Sakhalin 120
Amoco 120, 134-135
Royal Dutch Shell 120
Siberia 120, 134-135
Amoco 120, 134-135
Royal Dutch Shell 120
organization of and changes in industry 131-134
pipelines 141
Druzhba 141, 169
Turkey 141
political influence on business deals 134-135
Rule 2-9 135
pollution 168, 169
refineries 127-129, 159, 168, 169
European Russian 168
failed refinery deal 127-129
Russian 159, 169
Volga-Ural Refineries 169
reserves 125
specialists 86
Orlov, V. (Chairman of State Geological Committee) 125

Orujev (Deputy Minister of Oil) 91
Ostankino (Moscow TV station) 144
Ostashvily, A. (Chairman of the Russian National-Patriotic Front, *Pamyat)* 94
Paleolog, Sofia (daughter of last Byzantine emperor) 177
Parliaments 125, 129, 154-155
  conflicts with B. Yeltsin 154-155
  Russian Supreme Soviet 154-155
  Saratov oil industry 154-155
population 2, 3, 6, 7, 9, 12, 19, 20 40 , 41, 93-94
  Huns 3, 6
  Kazakhs 3, 6, 20
  Kirgiz 2, 7, 9, 12, 19, 40
  Mongols 3, 6
  Russians 3, 7
  Scythians 3, 6
  Tajiks 47
  Tatars 3, 6
  Turki 3, 6
  Uzbeks 7, 9, 19, 41
presidents 140-145
  elections 140-141
  involvement in business 140-145
property conflicts with Germany 152
Pugachev, E. (Cossack leader) 151, 53
relations with Tatarstan 103-104
religion 145
results of Russian education program 77-81
Ribkin, I. (Chairman of Russian State Duma) 174
Rizjkov (Prime Minister) 180-181
  Uspenskaya, N. (Economics Advisor) 180-181
Rostov 177
Rutskoy, A. (former Vice-President) 28, 139, 140
Saratov (see *Saratov)*
Saratovneftegeofizika (Saratov Oil Geophysical Survey) 18, 147-149, 150, 155
  Michurin, A. (President) 149
  modernization 147
  privatization 147-149
    vouchers in FSU businesses 148-149
  Zakharov, P. (Vice-President) 149-150, 155
Sayidov (Chief of Dagestan Oil) 89-90
Sevastopol 177
Shafranik, Y. (Minister of Fuel and Energy) 161
Shevardnadze, E. (FSU Foreign Minister, Georgia's KGB Chief, First Secretary of Communist Party) 75, 84
  coup d'etat in Georgia 75
shipping problems 163-164, 166
Shnurnov, V. (President of KEDR) 130
Shumeyko, V. (Deputy Prime Minister, Speaker for the Chamber of Parliament, Vice-Premier Minister) 28, 135
Siberia (see *Siberia)*
socialistic nations 178
Stalin, J. 47-48, 72, 73, 76, 93-94, 178
  creation of republics 178
  education 76
  Ossetia 76
  Tatarstan 93-94
Stalin, Svetlana Alilueva (J. Stalin's daughter) 145
standards of exploration 57
State Investment Committee 129
Stolypin (Prime Minister) 144
Suvorov, Generalissimus A. 151, 152
Tabakova, E. (Chief of Technical Department's Port Authority of Tuapse) 179
Taslitsky, E. (Deputy Minister of Fuel and Energy) 168
Tatarstan (see *Tatarstan)*
theft of resources 180-181, 193
  government intervention 181
  Rule 2-23 181, 193
Tolstoy, M., Ph.D. (member of the Russian Parliament and of State Duma) 172
travel problems 163-164

trade and investment 119-120, 1130-131, 134-135, 170, 191
joint ventures with the U.S. 119-120
laws concerning 130-131
Rule 2-6 130
political influences 134-135
Rule 2-21 170, 191
Trans-Siberian Railway 159
Udod, V. (Deputy Director of Regional State Tax Management branch in Tyumen) 174
Urazjtcev (member of first Russian Parliament) 88-89
Uspenskaya, N. (Economics Advisor) 180-181
Rizjkov (Prime Minister) 180-181
Vasiliy III (Grand Prince of Moscow) 177
Volga River 18
Volgograd (formerly Stalingrad) 167
Western Siberia (see *Western Siberia)*
Yavlinksy (Economic Advisor, Member of the Parliament) 125, 126
Yeltsin, B. (Chairman of the Supreme Soviet of the Russian Federation) 73, 89, 95, 97, 121, 124, 125, 128, 129, 132-133, 136-138, 142, 145, 152, 154-155, 173, 174, 177
Alekperov, Vagit (Lukoil President, Deputy Oil Minister Minister) 142
Burburlis (Yeltsin associate) 128
condition of government 121
conflicts with Parliament 154-155
democracy 136-138
Rule 2-10 138
Rule 2-11 138, 192
property conflicts with Germany 152
gold 173, 174
Golovaty, A. (Deputy Minister of Finance) 173
secret resolution creating state monopoly of reserves 173, 174
Udod, V. (Deputy Director of Regional State Tax Management branch in Tyumen) 174
Tatarstan oil agreement 97
Yermolov (Russian General) 72
Ziuganov, G. (Leader of the Russian Communist Party) 133
Zjirinovsky (presidential candidate) 145
Zolotukhin, V. (Member of Parliament) 48-49
France 7, 16
Frankenburgh (Liechtenstein) 142

G

Gabrielyants, G. A. (FSU Minister of Geology) 132
Gaidar, Y. (acting Russian Prime Minister, President of Astrakhan oil company) 134, 139, 160
inflation 139
Sacks, J, (former Harvard adviser to the Russian government) 139
Galburaev, Ruslan (former Chief Engineer of Sakhalin Oil, Minister of Chechen Republic's Oil Industry) 74, 80, 83, 84-86, 91
Gamsakhurdia (President of Georgia) 72, 73, 74, 75, 76, 80, 86, 88, 91
gas, natural 48, 50, 52, 67
recoverable reserves (Turkmenistan) 67
Genesis 76
quote from 76
geological survey 14
Georgia 74, 75, 76, 84, 140, 178, 179, 180
Abkhazia 75, 76
independence 75
Batumy, port of 74, 84
Babilashvily, A. (Professor of History, University of Tbilisi) 76
Black Sea 179
Communism 75, 84

Gamsakhurdia (President) 72, 73, 74, 75, 76, 80, 86, 88, 91
history 178
KGB 75, 84
nationalism 142
Ossetia 76
Stalin 76
president 140
Shevardnadze, E. (FSU Foreign Minister, Georgia's KGB Chief, First Secretary of Communist Party) 75, 84
coup d'etat in Georgia 75
Tamara (Princess) 76
Tbilisi (capital) 75
theft of oil 180
geosynclinal formation 14
Germany 12, 72, 73, 87, 123, 151, 152, 178
Berlin 87
history 151, 152
property conflicts with Russia 152
Yeltsin, B. 152
von Bismarck, O. (Chancellor of the German Empire) 123
World War II 72, 73
gifts (Rule 1-16) 39, 108
Godunov, Boris Tsar 144-145
gold
Almalyk refining plant (Uzbekistan) 58
Angren Mill (Uzbekistan) 58
BMB (Turkish gold company) 55
deposits
Kirgizstan
Kum-Tor 7, 14, 15, 31, 32, 38, 44, 45, 46, 87
contract 87
Uzbekistan
Akgol 58
Jalair 58
Karakutan 58
Kattaich 58
Kayragash 58
Keskan 58
Koch Bulak 58
Kyzyl Almasay 58
mining method 60
Kyzyl-Tor (Kirgizstan) 14, 15, 38
Mezjdurchye 58
Muruntau 53
Samarchuk 58
Samich 58
Sop 58
Temirkabuk 58
Uchkulach 58
Yakhton 58
Zarmitan 55, 57
description of 57
heap leaching technology 56
Hoshars (party for neighbors, Uzbekistan) 63-66
conducting business at 63-66
London's gold market 62
maps
eastern Uzbekistan's deposits 59
Uzbekistan's deposits 54
mineral resources 58
Margjanbulack Enriching Plant (Uzbekistan) 57
Newmont (North American gold company) 17, 29, 37, 39, 53, 55, 64
Cambre, M. (CEO) 55
European Development Bank 55
privatization of Uzbekistan's industry 62
refining plants, 58
Russian reserves 125
South African reserves 125
Turkey
BMB (gold company) 55
underground mining 58
Uzbekistan reserves 58
Yeltsin, B. 173, 174
Golovaty, A. (Deputy Minister of Finance) 173
secret resolution creating state monopoly of reserves 173, 174
Udod, V. (Deputy Director of Regional State Tax Management branch in Tyumen) 174
Golovaty, A. (Deputy Minister of Finance) 173

Gorbachev, M. (General Secretary of the Communist Party) 48, 73, 75, 87, 103-104, 124, 133, 136, 147, 148-149, 177
  coup d'etat against (1991) 103-140
  privatization of businesses 147, 148-149
    vouchers in FSU businesses 148-149
  Tengiz oil contract with Chevron 87
Gorbachev, M. (former Deputy Minister of Interior Affairs and L. Brezhnev's son-in-law) 52-53
Gore, Al 8
Gosplan (Soviet State Planning Committee) 79
  Baybakov (Chairman, Oil Minister, Prime Minister) 79
gray cardinal (influential person in Russian affairs) 128
Greece 42
greeting citizens of the FSU (Rule 1-9) 33, 107
Grozneftegeofizika (Chechen Republic oil company) 82, 149
  Peru contracts 149
Grozny (capital of the Chechen Republic) 72, 73, 74, 75, 80, 81, 86, 88, 91, 113
  apartments 113
  crime 73, 113
  dachas 113
  Eastern Region oil fields 81
  German army 72
  hotels 113
  Nobel's oil fields 86

H

Harvard 139
  Sacks, J, (former adviser to the Russian government) 139
Harvey, J. (DLJ Securities Analyst for Amoco) 174
healthcare (Uzbekistan) 52
heap leaching technology (gold) 56
HEMCO—Hunt Exploration and Mining Co. 17, 32
  Hunt, Nelson B. (President) 32
Herzen, A. I. (Russian writer) 130
Hoshars (party for neighbors, Uzbekistan) 63-66
  Rule 1-25 64, 112
  Rule 1-26 64, 108
hospitality 71-72, 150, 151, 153, 193-4
  Hoshars 63-66
    Rule 1-25 64, 112
    Rule 1-26 64, 108
  in the Chechen Republic 71-72
  Rule 2-14 150, 194
  Rule 2-15 151, 194
  Rule 2-16 151, 193
  Rule 2-17 153, 193
  Rule 2-18 153, 193
hotels 113
  Intourist 113
Hunt, Nelson B. (President, HEMCO—Hunt Exploration and Mining Co.) 32
Hutson, T. R. (U.S. Embassy First Secretary) 6

I

Idinov, K. (Chairman of Standing Commission for International Relations, Kirgizstan) 46
Irkutsk 162, 163
India 149
Ingush 72, 73, 78, 89
  Chechen-Ingush Autonomous Republic 72, 78
  Chechen Republic, the 72
  population 72-73
Interior Affairs Ministry 153
  crime 153
  mafia 153
International Monetary Fund 175
international relations, discussing (Rule 1-29) 71, 109
interpreter, hiring a local (Rule 1-17) 41, 112
Intersystems (Uzbekistan) 48-49
  Solotukhin (President) 48-49
Intourist 30, 113
  hotels 113
Institute of Combustible Fossils (IGIRGI of the FSU) 79
introducing yourself and your company 105-106
  Rule 1-14 105-106

Rule 1-23 106
Rule 1-37 105
Rule 1-38 106
Rule 1-39 106
IRA (U.S. oil company) 85
Iran 7, 55, 67
Rafsanjany, (President) 7
railroad to Turkmenistan 67
Islam 3, 9, 47, 71, 74, 102
fundamentalists 48
Ishmaelites 47
Koran 6
Muslims 3, 9, 47, 71, 74
Prophet 71
questioning religious beliefs 102
Shiites 47, 71
Sunnis 9
Issykkul region (Kirgizstan) 44
Israel 28, 29, 78, 87
Italy 7
Ivan the Terrible 77, 93, 144-145
Chechen society 77
conquest of Tatarstan 93

J

Jabrailov (Chief Engineer of Dagestan Oil) 89
Japan 7, 16, 64, 166
Uzbekistan refining project 64
Jewish autonomous region (Russian Far East) 120
Jews 47, 73, 78, 93, 177
freedom 177
Heroes of the Soviet Union 93
influence on education 77-81
in the Chechen Republic 73
Johnson, Lyndon B. 81
affirmative action programs 81
joint ventures 164, 170-171, 189, 191
four documents required 170
registering in FSU 170
Rule 2-21 170, 191
Rule 2-22 171, 189
Russian government lists for 164

K

Kaganovich, L. (First Secretary in Kiev) 94
Kalmyk 78
education 78
Kara-Kum desert
recoverable natural gas reserves (Turkmenistan) 67
Kara Sea 175
offshore discoveries 175
Karimov (President of Uzbekistan) 61
Karpov, A. (World Chess Champion) 127-129
Kazan (capital of Tatarstan) 93, 96, 98, 99, 102, 113
accommodations 113
Kazakhstan
Almaty (capital) 6, 113
accommodations 113
Chechens 72
banishment to Kazakhstan 72
education 78
Karachaganck oil field 154
Mangyshlack Peninsula 20
Nazarbaev, N. (President) 6, 19, 66
Pre-Caspian Depression 20, 82, 153-154
Tengiz oil field 20, 154
theft of resources 181
World War II 152
Kennedy, John F. 81
affirmative action programs 81
KGB 28, 30, 48, 52, 73, 75, 84, 133, 137, 152-153, 193
Andropov, Yuri (former KGB head, General Secretary of the Soviet Union's Communist Party) 52
annihilation of Cossacks 73
Gamasakhurdia (President of Georgia) 75
private security services 152-153, 193
Rule 2-17 153, 193
Rule 2-18 153, 193
Shevardnadze, E. (FSU Foreign Minister, Georgia's KGB Chief, First Secretary of Communist Party) 75, 84
coup d'etat in Georgia 75
Khackulov (Deputy Minister of Uzbekistan) 56

Khasbulatov, Rouslan (Speaker of first Russian Parliament) 88-89
Kherson 177
Kholbaev, T. (General Director of Kirgizstan's state oil company) 19, 22-28, 40-43
Khomenuk, L. (President of Kum-Tor Operations Company, Cameco's branch) 44, 46
Khackulov (Deputy Prime Minister, Chairman of *Uzbekeftegaz)* 56, 63
Khandiza polymetallic deposit (Uzbekistan) 56
Kiev 94
  Kaganovich (First Secretary) 94
Killborn Western, Inc. 45
Kirgizstan
  Akaev, Askar (President) 13, 14, 19, 28, 29, 32, 43-44, 45, 46, 87
  Alay depression 18
  Aytmatov, Chingez (writer) 23
  Birshtine, B. (associate of President Akaev) 32
  Bishkek (capital) 12, 13, 17, 19, 22, 29, 30, 32, 33, 35, 40, 46
  Bishkek Business School 44
  Chatkal area 13, 16, 31
  Chengyshev (Prime-Minister) 40-43, 87
    Chevron 20, 43, 87
      Tengiz oil contract 87
  Chu depression 18
  Chu Region, The 45
    Zjumagulov, A. (Governor) 45
  culture 12-13
  education 78, 81
  Fergana intermountain basin 18, 19, 21, 23, 41, 43
  geographical features, description of 14
  Idinov, K. (Chairman, Standing Committee for International Relations of Parliament) 46
  Issykkul region 44
  Jalal University 44
  Jeruy gold deposit 38
  Kara-Kim desert 67
  Kholbaev, T. (General Director of Kirgizstan's state oil company) 19, 22-28, 40-43
  Khomenuk, L. (President of Kum-Tor Operations Company, Cameco's branch) 44, 46
  Kirggeology (State Committee of Geological Exploration) 30
    Tekenov, S. (General Director) 19, 22, 30, 31, 40-43
  Kirgisneft (state oil company) 19, 22-28, 40-43
    Kholbaev, T. (General Director) 19, 22-28, 40-43
  Kirgizs 16, 17, 23
  Kirgizaltyn (state gold mining company) 44
    Sarygulov, Dastan I. (General Director) 44
  Kirgizzoloto (gold mining company) 31, 35-38, 39, 40-43
    Kydrov, Kapar K. (General Director) 35-38
    Serebryansky, Anatoly T. (Deputy Director) 35, 40-43
  Kisziz region 44
  Kulov, F. (Vice-President) 8, 17, 32, 35
  Kum-Tor gold deposit 7, 14, 15, 31, 32, 38, 44, 45, 46, 87
    contract 87
  Kydrov, Kapar K. (General Director of *Kirgizzoloto)* 35-38
  Kyzyl-Tor gold deposit 14, 15, 38
  Levitin, L. (President's State Adviser) 43-44
  mineral resources 10-11, 14-15
  Ministry of Energy and Fuel 25
  neighbors 9
  oil industry 149
  Oil Ministry in Moscow 96
  Omuraliev (Minister of Industry) 25, 40-43
  Osh region 14
  Osmonov, B. (Member of Parliament) 13
  Parishkura, M. (Deputy Minister of Foreign Trade) 33
  periodicals
    Molodezj Kirgizstana 21-22
    Slovo Kirgizstana 32, 44, 46
  population 9, 12

Sarygulov, Dastan I. (former Governor of Talas region) 44
Shamshiev, Bolot T. (Member of Parliament) 12, 23, 33, 34
Talas region 44
Sarygulov, Dastan I. (former Governor) 44
Taldy-Bulak gold deposit 14, 32, 38
Tengiz oil field
Chevron 20, 43, 87
Tien Shan mountains 9
Zjelal-Abad region 41, 42
Zjeruy gold field 16, 31, 32, 38
Zjumagulov, A. (Governor of the Chu Region) 45
Kish, W. (Seagram Manager) 182
Kissinger, H. (former U.S. Secretary of State) 122
Komi Autonomous Republic (Northern European Russia) 120, 135, 155
Texaco oil project 120, 135
Kononova, M. (Editor of *Uzbekistan-Contact)* 56
Koran 6
Korchinsky, D. (Deputy of Ukraine's National Assembly) 179
Krasnodar (Chechen Republic) 81
Western Region oil fields 81
Krasnovodsk oil refinery (Turkmenistan) 66
Krasnoyarsk region (Siberia) 120
Khrushchev, M. (N. Khrushchev's daughter) 179
Khrushchev, Nikita 72, 79, 177, 178-179
Gosplan (Russian State Planning Committee) 79
Baybakov (Chairman, Oil Minister, Prime Minister) 79
nuclear war 179
Krylov, I. (Chief of Staff of Representatives of Russian President) 152
Kulov, F. (Vice-President of Kirgizstan) 8, 17, 32, 35
Kuwait 21
Kuznetsov, V. (General Director of *Yuzjmorneftegeofizika,* Southern Marine Geophysics) 187
Kuznetsov (First Deputy Chairman of the Byelorus Republic) 143
Kydrov, Kapar K. (General Director of *Kirgizzoloto)* 35-38
Kyzyl-Kum desert 47, 49, 53

L

languages 47, 72
Laws, R. (Representative of American Federal Reserve System to the Byelorus Republic) 143
lead 14, 57
Lenin, V. I. 53, 77, 123
Levitin, L. (State Adviser to President of Kirgizstan) 43-44
Libya 125
Liechtenstein 142
Frankenburgh (oil company) 142
Lithuania 28, 93
Brazauskas, A. (President) 28
Lisovsky (Head of Geological Management of FSU Oil Minister) 89-90
Lloyd's of London 128
London 62, 87, 128
gold market 62
Lloyd's of London 128
Lukoil (Russian company) 141-142
Azerbaijan oil deal 141-142
Frankenburgh lawsuit 142
Mitcui loan 142
World Bank loan 142
world investments 142

M

Macedonia 42
mafia 84, 91, 153, 180
theft of oil 180
Major, John (British Prime Minister) 141
Azerbaijan oil deal 141
Mangyshlack Peninsula 20
Marx, K. 123
Mary region 120
Matlock, J. (former U.S. Ambassador to FSU) 33
McNeill, K. P. (Vice-President of MD SEIS, INC.) 82

meals (includes author's recommendations) 114
meetings, business 38, 41
  attending Hoshars (party for neighbors, Uzbekistan) 63-66
    Rule 1-25 64, 112
    Rule 1-26 64, 108
    Rule 2-16 151
  conducting business meetings 38, 41
    Rule 1-13 38, 110
    Rule 1-18 41, 107
    Rule 1-19 42, 107
    Rule 1-20 42, 109
    Rule 1-23 61, 106
mercury 14
Meshkov, J. (Crimean President) 188
Mesozoic age 14
Mesozoic deposits 157
Mesozoic sandstones 157
Michigan University's Center of Study of the U.S.S.R. and East Europe 140
Michurin, A. (President of Saratov Oil Geophysical Survey) 149
Minbulak region 21, 24, 50, 53, 62-63
  Kirgizstan 21, 24
  oil drilling 50
  oil-well explosion 62-63
  Uzbekistan 53, 62
Minproc and Chilewich 17
Mitcui loan to Lukoil 142
Mokashov, A. (General of Soviet Army) 144
Moldavia 47, 142
  nationalism 142
Moldova 28, 94
  population 94
  Snegur, M. (President) 28
Molodezj Kirgizstana (periodical) 21-22
money, concept of 123-125, 138, 192-193
  Rule 2-1 123, 192
  Rule 2-11 138, 192
  Rule 2-23 193
Mongolia 20
Morland, A. (Vice-President of Amoco) 120
Moscow
  all information available 163
  American investments in 120
  Belov, V. (President of refinery) 127-128
    Alexander (employee) 127-128
    Aliev, Geydar (KGB Chief, First Secretary of Azerbaijan's Communist Party) 141
    failed deal 127-128
    Protcenko (Vice-President of refinery; former Chief of Moscow Regional Press Club) 127-128
  buying food from Ukraine 187
  Communism and Socialism 123, 136
    Committee of the Communist Party 136
      Yeltsin, B. 136
    Marx, K. 123
    von Bismarck, O. (Chancellor of the German Empire) 123
  consulting when starting a business (Rule 1-36) 97, 110
  cost of living 139
  crime 138-140
  false oil reports 158
  Federal Express 114
  Gorbachev, M. 48, 73, 75, 87, 103-104, 124, 133, 136, 147, 148-149, 177
    coup d'etat against (1991) 103-140
    privatization of businesses 147, 148-149
      vouchers in FSU businesses 148-149
    Tengiz oil contract with Chevron 87
  Ministry of Fuel and Energy 133, 134
  Ministry of Geology 30, 131, 132, 134
    Gabrielyants, G. A. (Minister) 132
  Ministry of Oil 18, 26, 96, 131, 133, 134
  Ministry of the Gas Industry 133, 134

Chernomyrdin (Minister, Prime Minister) 133
Moscow Cinema House 144
Nezavisimaya Gazeta (Moscow daily) *139*
Tretyakov, V. (Chief Editor) 139
Ostankino (Moscow TV station) 144
attack on 144
Petroleum Institute 80
political influence on oil industry in FSU 159
Sheremetyevo (Moscow International Airport) 127
Slavyanskara (hotel) 127
Soyuzalmazzoloto (Union's Diamonds and Gold enterprise) 16
Stalin, J. 47-48, 72, 73, 76, 93-94, 178
creation of republics 178
education 76
Ossetia 76
Tatarstan 93-94
Vasiliy III (Grand Prince of Moscow) 177
Moslem's Belt (see the *FSU)*
Moslem nations of the North Caucasus 94
Muslims 3, 9, 47, 71
fundamentalists 48
Ishmaelites 47
Muslims 9
Shiites 47, 71
Sunnis 9
Mutalibov (President of Azerbaijan) 140, 141-142

N

Naberezjnie-Chelny (formerly the city of Brezhnev) 97
Najimov, S. (Chairman of State Committee for Precious Metals of Uzbekistan) 61, 64
Napoleon 155
National Westminster Bank of Scotland (NatWest) 56
natural gas 67
Turkmenistan 67
Naydenov, V. N. (Chief Geologist of Black Sea Oil and Gas) 187
Nazarbaev, N. (President of Kazakhstan) 6, 19, 66
negotiating contracts 38, 42, 61, 64, 84-86, 87, 88-89, 99-102, 153, 167, 170, 171, 181, 189, 191, 193
protocols 101
Rule 1-12 38, 112
Rule 1-13 38, 110
Rule 1-19 42, 107
Rule 1-20 42, 109
Rule 1-21 111
Rule 1-23 61, 106
Rule 1-25 64, 112
Rule 1-26 64, 108
Rule 1-33 87, 109
Rule 1-34 87, 111
Rule 1-39 101, 106
Rule 2-18 153, 193
Rule 2-20 167, 191
Rule 2-21 170, 191
Rule 2-22 171, 189
Rule 2-23 181, 193
Neogenic deposits, Lower (Chechen Republic) 83
neogenic sequences 18, 49
Newmont (North American gold company) 17, 29, 37, 39, 53, 55, 64
Cambre, M. (CEO) 55
European Development Bank 55
Nezavisimaya Gazeta (Moscow daily) 139
Tretyakov, V. (Chief Editor) 139
Nigeria 142
oil fields 142
Nikosia (Cyprus) 87
Nixon, Richard M. 171, 188
Niyazov (President of Turkmenistan) 66
Nobel, Alfred 86
oil fields in Grozny 86
Norsk Hydro 135
Timan-Pechora (American-owned Russian oil company) 135
North Caucasian Petroleum Research Institute 73-74, 84-86
Sokolovsky, Professor E. V. (Director) 73-74, 84-86

Novorossisk, port of 20, 84, 167, 177, 179, 180
Novorussia (New Russia) 177
nuclear bombs 77

O

Odessa 177
oil industry
  Azov, Sea of 181, 183-187
    oil and gas fields 183-187
      map of 186
  Black Sea 177-188
    map of offshore areas 185
    oil and gas fields 183, 184
      East Severskoe 183
      Krasnodar 183
      North Chernomorsky Rift 184
      offshore 184
      Pre-Caucasian Basin 183
      Severskie 183
    ports of 43, 74, 84, 141, 167, 169, 179, 180
      Abkhazia 76
      Batum 74, 84
      Novorossisk 20, 84, 167, 177, 179, 180
      Tuapse 179, 180
  Bokhara-Hivinsky basin (Uzbekistan) 50
  Byelorus Republic 143
    Byelorusneftegeofizika 149
  Chechen Republic industry (see *the Chechen Republic)*
  Crimean oil and gas production 182, 184
  drilling in Uzbekistan 50
  Druzhba (Russian pipeline) 141, 169
  exploration 87, 166, 175
    Rule 2-19 166, 191
    with submarines 87, 175
  Fergana intermountain basin 18, 19, 21, 23, 41, 43
    Minbulak area 21, 24
  foreign investments 161-162, 170-174, 189, 191, 193
    government programs 161-162
    joint ventures 161, 170-171, 193
      Polar Lights 161
      Rule 2-21 170, 191
      Rule 2-22 171, 189
      Rule 2-23 181, 193
      White Nights 161
    laws concerning 172-174
    Parliament (Soviet and Russian) 172-173
    Yeltsin 173, 174
  Galburaev, Ruslan (Minister of Chechen Republic's Oil Industry) 74, 80, 83, 84-86, 91
  government involvement with 96
  Grozneftegeofizika (Chechen Republic oil company) 82, 149
    Peru contracts 149
  Gumkhana field (Uzbekistan) 50
  Hoshars (party for neighbors) 63-66
    conducting business during 63-66
  Japanese refining project (Uzbekistan) 64
  Kara Sea 175
    offshore discoveries 175
  Karshy basin (Uzbekistan) 50
  Kirgizstan industry (see *Kirgizstan)*
  Kubarnalakma oil field 50
  legislation concerning industry 172-174
    export taxes 174
  long-term issues 175-176
  Mangyshlack Peninsula 20, 49
    blowout preventers (BPOs) 49
    Kazakhstan 20
    sulfur content 49
  Nobel, Alfred 86
    oil fields in Grozny 86
  oil fields 66-67, 142
    Egypt 142
    Nigeria 142
    eastern Caspian Sea (Turkmenistan) 66-67
    Tobago 142
    Trinidad 142
    Tunis 142
  Oil Ministry in Moscow 96
  Orujev (FSU Deputy Minister of Oil) 91
  pipelines 20, 21, 43, 49, 50, 74, 141

Persian Gulf 67
Pre-Caspian Depression 20, 82, 153-154
railroad between Turkmenistan and Iran 67
refineries 20, 43, 50, 66, 73, 87, 95, 99-102, 127-129, 159, 168, 169
  Chardzjou (Turkmenistan) 66
  Chechen Republic, the 73
  Russian 127-129, 159, 168, 169
    European Russian 168
      Moscow 168
      Nizhny 168
      Novgorod 168
      Ryazan 168
      Yaroslav 168
    failed refinery deal 127-129
    Russian 159, 169
      Achinsk 159
      Omsk 159, 169
      Tomsk 159, 169
    Volga-Ural Refineries 169
      Ishimbay 169
      Samara 169
      Saratov 169
      Syzran 169
  Tatarstan 95, 99-102
  Tengiz oil field (Kirgizstan) 20, 43, 87
    Chevron contract 87
  Ust-Yurt basin (Uzbekistan) 50
Russian industry (see *the FSU* and *Russia)*
Sakhalinneft (Sakhalin Oil) 80
Saratov industry (see *Saratov)*
Schlumberger's equipment 84
seismic surveys 82
  Bally (Professor, Rice University) 82
solid bitumen 95-96, 104
Tatarstan industry (see *Tatarstan)*
theft of oil 180-181, 193
  Rule 2-23 (contract only for oil amounts and prices on board ships, FOB) 181, 193
Ukraine industry (see *Ukraine)*
Western Siberia industry (see *Western Siberia)*
oil-gas condensate field 7
Omsk region (Siberia) 120
Omuraliev (Kirgizstan Minister of Industry) 25, 40-43
Orlov, V. (Chairman of State Geological Committee) 125
orogenic formations 14
Orujev (FSU Deputy Minister of Oil) 91
Osmonov, B. (Member of Kirgizstan Parliament) 13
Ossetia 76
Ostankino (Moscow TV station) 144
  attack on 144
Ostashvily, A. (Chairman of the Russian National-Patriotic Front, *Pamyat)* 94

P

Paine Webber Inc. 96
  Sadic-Khan, Orhan (Managing Director) 96
Paleogenic terrigenous sequences 18
Paleolog, Sofia (daughter of last Byzantine emperor) 177
Pamyat (Russian National-Patriotic Front) 94
  Ostashvily, A. (Chairman) 94
Pan-Turkism 48
Paraguay 149
Paris 150
Parishkura, M. (Kirgizstan Deputy Minister of Foreign Trade) 33
paying contractual fees (Rule 1-21) 44, 111
paying local expenses (Rule 1-15) 39, 108
periodicals 32, 44, 46
  Uzbekistan-Contact 56
    Kononova, M. (Editor) 56
  Molodezj Kirgizstana 21-22
  Nezavisimaya Gazeta (Moscow daily) 139
    Tretyakov, V. (Chief Editor) 139
  Russkaya Rech 179
  Slovo Kirgizstana (Kirgizstan) 32, 44, 46
perestroyka 52, 177
Persian Gulf 67

personal relationships 23, 26, 31, 33, 35, 106-107
advantages of 106-107
Rule 1-1 106
Rule 1-2 106
Rule 1-24 107
Rule 1-6 31, 107
Rule 1-8 33, 111
Rule 1-10 35, 107
Rule 1-11 35
Peru 149
Byelorusneftegeofizika contract 149
Petron, Inc. 29, 81
Buckley, P. (Vice-President) 29, 81
phone service in the FSU 114
phosphate 125
physicians 79
Rule 1-30 79, 112
Poland 93, 178
political conditions in the Chechen Republic 73-76
Sokolovsky, Professor E. V. (Director of North Caucasian Petroleum Research Institute) 73-74
U.S. involvement 75-76
Ames, (CIA Officer and Russian Spy) 75
political leaders, discussing (Rule 1-6) 32, 107
politics, discussing (Rule 1-5) 30
pollution 52-53
Aral Sea 52
Gorbachev, M. (former Deputy Minister of Interior Affairs and Brezhnev's son-in-law) 52-53
population 9, 12, 37, 72, 73, 93-94
the Chechen Republic 72
Ingush 72-73
Russian 37, 72, 73, 93-94
Tatarstan 93-94
postal services in the FSU 114
privatization of FSU businesses 147, 148-149
vouchers in FSU businesses 148-149
privatization of Uzbekistan's gold industry 61-62
prostitution 67-68, 112, 113, 180
Rule 1-28 68, 112
Protcenko (Vice-President of refinery;Chief of Moscow Regional Press Club) 127-128
failed refinery deal 127-128
protocols in business 101
protocol of intentions to form a joint venture 170-171, 191
Rule 2-21 170, 191
Pugachev, A. (Cossack leader) 151, 152
punctuality of local VIPs (Rule 1-35) 96, 107

R

Rakhimov, A. (First Deputy of Uzbekistan, Chairman of Uzbek Oil and Gas) 62-63, 64
Rafsanjany, (President of Iran) 7
railroads 67
between Iran and Turkmenistan 67
Uzbekistan 56-57
Denau train station 56
Uzuh train station 57
Rashidov, S. (First Secretary of Central Committee of Uzbekistan's Communist Party) 52-53
Regional Mining Society 36
Religion 3, 6, 9, 47, 48, 71, 74, 93, 102
Catholicism 93
Christianity 71
Islam 3, 6, 9, 47, 48, 71, 74, 102
fundamentalists 48
Ishmaelites 47
Koran 6
Muslims 9
Prophet 71
Shiites 47, 71
Sunnis 9
Orthodox Christianity 93
religious beliefs 102
and Communism 102
questioning 102
Republican Air Ambulance Service (FSU) 53
reputation, maintaining 189
Rule 2-5 189
Rule 2-22 189
restaurants (author's recommendations) 114

Reznichek, Antony (Russian interpreter) 129
Ribkin, I. (Chairman of Russian State Duma) 174
Rice University 82
Bally (Professor) 82
seismic surveys 82
Rijanov, A. (Chief gas and oil expert of Tatarstan Cabinet of Ministers) 96
Rizjkov (Soviet Prime Minister) 180-181
Uspenskaya, N. (Economics Advisor) 180-181
Roman Empire 12
Romania 179
Romer, Roy (Governor of Colorado) 64, 84
Rostov 177
Royal Dutch Shell 120
Rules for western business people 6-7, 12-13, 23-44, 61, 63-66, 67-69, 71-72, 79-80, 84-86, 87, 96, 97, 98, 99, 101, 102-103, 105-115, 123, 126, 127, 129, 130, 132, 134, 135, 138, 140, 150, 151, 153, 166, 167, 170, 171, 181, 189-194
1-1 (value of personal relationships) 23, 106
1-2 (using personal relationships in conducting business) 26, 106
1-3 (negotiating as a team with local representatives) 27, 110
1-4 (inviting partners to visit your local country) 28, 108
1-5 (discussing politics) 30
1-6 (willingness to be a pupil) 31, 107
1-7 (discussing political leaders) 32, 110
1-8 (working with clans) 33, 111
1-9 (greeting citizens of the FSU) 33, 107
1-10 (personal relationships and etiquette) 35, 107
1-11 (discussing wives and daughters) 35
1-12 (facing resistance in discussions) 38, 112
1-13 (setting followup meetings) 38, 110
1-14 (expressing wealth) 39, 105-106
1-15 (dealing with local expenses) 39, 108
1-16 (giving gifts) 39, 108
1-17 (hiring local interpreters) 41, 112
1-18 (beginning a business meeting) 41, 107
1-19 (listening as a part of negotiations) 42, 107
1-20 (chatting as a part of negotiations) 42, 109
1-21 (paying contractual fees) 44, 111
1-22 (traveling in the FSU) 53, 112
1-23 (willingness to help) 61, 106
1-24 (understanding your local partner) 63, 107
1-25 (including local agents at *Hoshars)* 64, 112
1-26 (excluding women from *Hoshars)* 64, 108
1-27 (firing local agents) 67, 112
1-28 (prostitution) 68, 112
1-29 (discussing international relations) 71, 109
1-30 (visiting a physician in the republics outside Russia) 79, 112
1-31 (appointing a local national as a specialist) 79, 111
1-32 (understanding the educational background of a local partner) 79, 111
1-33 (reliability of contracts) 87, 109
1-34 (ease of contract violation in the FSU) 87, 111
1-35 (punctuality of local VIPs) 96, 107
1-36 (starting a business) 97, 110
1-37 (importance of introductions) 98, 105
1-38 (supplying advertising about your company) 99, 106

1-39 (supplying documents to be signed) 101, 106
1-40 (payment of local partners) 102, 113
2-1 (using proper financial terms) 123
2-2 (contact with quasi-state enterprises) 126, 190
2-3 (when to contact with private enterprises) 127, 190
2-4 (when to avoid private enterprises) 127, 190
2-5 (moral and ethical standards) 129, 189
2-6 (following Russian laws and regulations) 130
2-7 (investigating oil fields) 132, 191
2-8 (avoiding Russian ministers) 134, 190
2-9 (avoiding political fights) 135
2-10 (appealing to the proper authority) 138
2-11 (don't expect a fast money turnover) 138, 192
2-12 (checking origins of investments) 140, 191
2-13 (handling conflicting opinions) 140
2-14 (screening for alcohol problems) 150, 194
2-15 (non-FSU employees with alcohol problems) 151, 194
2-16 (never refuse a drink) 151, 193
2-17 (hire private security companies) 153, 193
2-18 (avoid using cash) 153, 193
2-19 (don't hunt oil fields in far corners) 166, 191
2-20 (check your Russian partners' reports) 167, 191
2-21 (hire a Russian lawyer to prepare documents for joint ventures) 170, 191
2-22 (necessity of solid resources of time and money) 171, 189
2-23 (contract only for oil amounts and prices on board ships, FOB) 181, 193
Russkaya Rech (newspaper) 179
Russia
Adygey (Republic) 89
air force 77
Alexey, Tsar 178
Ames (CIA Officer and Russian Spy) 75-76
Arkhangelsk region 120
army 93, 133, 144
Mokashov, A. (General) 144
World War I 93
World War II 93
Astrakhan region 134-135
oil deal 134-135
Baybakov (Oil Minister, Prime Minister, Chairman of *Gosplan)* 79
Belogorod region 120
Beria 73
freed criminals 73
borders 75
Brezhnev, Leonid 52-53, 79
Gorbachev, M. (son-in-law, Deputy Minister of Interior Affairs) 52-53, 79
bribery 38-40, 102, 113
Rule 1-40 102, 113
Burburlis (B. Yeltsin's associate) 128
Catherine the Great (Ekatherine II) 151, 152, 177, 178
Pugachev, A. (Cossack leader) 151, 152
revolt against 151, 152
Suvorov, Generalissimus A. 151, 152
Chernomyrdin, Z. (Prime Minister, Minister of Gas Industry) 87, 133, 134, 135, 158, 175
International Monetary Fund 175
Civil War 177
Commonwealth of Independent States (CIS) 124
Communism 12, 16, 20, 28, 30, 47, 52, 57, 72, 73, 75, 102, 123, 171, 177, 178
Andropov, Yuri (KGB head, General Secretary of the FSU Communist Party) 52
annihilation of Cossacks, 73, 177

Armenia 47
Chechen-Ingush Autonomous Republic 72
concept of money 123
crime 75
Beria 73
Shevardnadze, E. (FSU Foreign Minister, Georgia's KGB Chief, First Secretary of Communist Party) 75
coup d'etat in Georgia 75
equality of citizens 178
Gorbachev, M. (General Secretary of the Communist Party) 136
Marx, K. 123
Moscow Committee of the Communist Party 136
Political Bureau of the Central Committee 136
religious beliefs 102
Rashidov, S. (First Secretary of Central Committee of Uzbekistan's Communist Party) 52-53
von Bismarck, O. (Chancellor of the German Empire) 123
Yeltsin, B. 136-138
democracy 136-138
Ziuganov, G. (Leader of Russian Communist Party) 133, 161
crime 38, 40, 73, 87-88, 113, 138-140, 150-151, 153, 167, 180-181, 193-194
alcohol-related problems 150-151, 193-194
crime 151
Rule 2-14 150, 194
Rule 2-15 151, 194
Rule 2-16 151, 193
bribery 38-40, 102, 113, 180-181
Rule 1-40 102, 113
Russian Coast Guard 181
fraud in ecological cleanups 167
Interior Affairs Ministry 153
mafia 84, 91, 153, 180
theft of oil 180
Moscow 138-140
prostitution 67-68, 112, 113, 180
Rule 1-28 68, 112
Dagestan 86, 87-88, 89
democracy 136-138, 144
Yeltsin, B. 136-138
Doudaev, Johar (President, Air Force General) 75, 77, 80, 87-88
nuclear bombs 77
Druzhba (Russian pipeline) 141, 169
economic and political situations 121-122, 139, 140
Gallup poll 121
reasons for difficult business situations 121
inflation 139
monetary policy 139
Rutskoy, A. (Russian Vice-President) 28, 139, 140
Ekatherine II (see *Catherine the Great*)
foreign investments 161-162, 170-174
government programs 161-162
joint ventures 161, 170-171
laws concerning 172-174
Parliament (Soviet and Russian) 172-173
Yeltsin, B. 173, 174
Gabrielyants, G. A. (Minister of Geology for the FSU) 132
Gaidar, Y. (acting Russian Prime Minister, President of Astrakhan oil company) 134, 139, 160
Sacks, J, (former Harvard adviser to Russia) 139
Godunov, Boris Tsar 144-145
gold reserves (see *gold*)
Golovaty, A. (Deputy Minister of Finance) 173
Gorbachev, M. (General Secretary of Communist Party) 48, 73, 75, 78, 87, 103-104, 124, 133, 136, 147, 148-149, 177
coup d'etat against (1991) 103-104
privatization of businesses 147, 148-149

vouchers in FSU businesses 148-149
Tengiz oil contract with Chevron 87
Gosplan (State Planning Committee) 79
Baybakov (Chairman, Oil Minister, Prime Minister) 79
government 71, 96
oil industry 96
Heroes of the Soviet Union 93-94
Herzen, A. I. (writer) 130
history 71, 72, 76, 93, 123-125, 144-145, 151, 152, 177-179
inflation 139
Gaidar, Y. (acting Russian Prime Minister, President of Astrakhan oil company) 139
Sacks, J, (former Harvard adviser to Russia) 139
Intourist 30, 113
hotels 113
Ivan the Terrible 77, 144-145
Jewish autonomous region (Russian Far East) 120
long-term issues for foreign investment 175-176
anti-Americanism 176
Kaganovich (First Secretary of Kiev) 94
Tatarstan 93-94
Karpov, A. (World Chess Champion) 127-129
KGB 28, 30, 48, 52, 73, 75, 84, 133, 137, 152-153, 193
Andropov, Yuri (KGB head, General Secretary of the FSU Communist Party) 52
coup d'etat in Georgia 75
private security services 152-153, 193
Rule 2-17 153, 193
Rule 2-18 153, 193
Zhevradnadze, E. (FSU Foreign Minister, Georgia's KGB Chief, First Secretary of Communist Party) 75, 84
Khasbulatov, Rouslan (Speaker of Russian Parliament) 88-89
Komi Autonomous Republic 120, 135, 155
Texaco oil project 120, 135
Krasnoyarsk region 120
Khrushchev, M. (N. Khrushchev's daughter) 179
Khrushchev, Nikita 72, 79, 177, 178-179
State Planning Committee 79
Baybakov (Chairman, Oil Minister, Prime Minister) 79
nuclear war 179
Krylov, I. (Chief of Staff of Representatives of Russian President) 152
Kuznetsov (First Deputy Chairman of Byelorus Republic) 143
Lenin, V. I. 53
mineral resources 125
Ministry of Ecology 167, 169
oil pollution 167
Ministry of Finance 173
Golovaty, A. (Deputy Minister) 173
Ministry of Fuel and Energy 133, 134, 161, 168
Shafranik, Y. (Minister) 161
Taslitsky, E. (Deputy Minister) 168
Ministry of Geology 30, 132, 131, 132, 134
Gabrielyants, G. A. (Minister) 132
Ministry of the Gas Industry 133, 134, 135, 148
Central Geophysical Expedition 148
Chernomyrdin, Z. (Minister, Prime Minister) 133, 134, 135, 158, 175
International Monetary Fund 175
Ministry of Oil 131, 133, 134, 148
sale of information 148
Minsk 91, 143
Mokashov, A. (General) 144
money, concept of 123-125, 192
history 123-125
Rule 2-1 123, 192
Moscow (see *Moscow*)

Moscow Petroleum Institute 80
Nezavisimaya Gazeta (Moscow daily) 139
Tretyakov, V. (Chief Editor) 139
Novorossisk, port of 20, 84, 167, 177, 179, 180
Novorussia (New Russia) 177
nationalism 142-145
natural gas 67
pipeline 67
nuclear bombs 77
oil industry
American-Russian joint ventures 120
areas where western investment is encouraged 162-168
aspects of industry 122, 123-135
conditions of industry 121-122
exploration 87, 166, 175, 191
Rule 2-19 166, 191
with submarines 87, 175
exports 120
government involvement 96
legislation concerning 172-174
export taxes 174
long-term issues 175-176
Lukoil (Russian company) 141-142
Azerbaijan oil deal 141-142
Frankenburgh lawsuit 142
Mitcui loan 142
World Bank loan 142
world investments 142
oil fields 120, 134-135, 141-142
Astrakhan 134-135
Azerbaijan 141-142
Apache oil deal 141-142
history 141
offshore oil fields 142
Azery 142
Gunashtly 142
Shirak 142
Komi 120
Texaco project 120
offshore zone 184
Roman Trebs 135
Sakhalin 120
Amoco 120
Royal Dutch Shell 120
Siberia 120, 134-135
Amoco 120, 134-135
Royal Dutch Shell 120
organization of and changes in industry 131-134
pipelines 141
Druzhba 141, 169
Turkey 141
political influence on business deals 134-135
Rule 2-9 135
refineries 127-129, 159, 168, 169
European Russian 168
Moscow 168
Nizhny 168
Novgorod 168
Ryazan 168
Yaroslav 168
failed refinery deal 127-129
Russian 159, 169
Achinsk 159
Omsk 159, 169
Tomsk 159, 169
Volga-Ural Refineries 169
Ishimbay 169
Samara 169
Saratov 169
Syzran 169
reserves 125
specialists 86
Oil Ministry 96
Orlov, V. (Chairman of State Geological Committee) 125
Orujev (Deputy Minister of Oil) 91
Ostankino (Moscow TV station) 144
attack on 144
Ostashvily, A. (Chairman, Russian National-Patriotic Front) 94
Parliaments 125, 129, 154-155
conflicts with B. Yeltsin 154-155
Russian Supreme Soviet 154-155
Political Bureau of the Central Committee 136

Yeltsin, B. 136
population 37, 72, 73, 93-94
presidents 140-145
elections 140-141
involvement in business 140-145
property conflicts with Germany 152
Protcenko (Vice-President of refinery, Chief of Moscow Regional Press Club) 127-128
Pugachev, A. (Cossack leader) 151, 152
relations with Tatarstan 103-104
religion 145
Ribkin, I. (Chairman of Russian State Duma) 174
Rizjkov (Prime Minister) 180-181
Uspenskaya, N. (Economics Advisor) 180-181
Rostov 177
Russian National-Patriotic Front, *Pamyat)* 94
Ostashvily, A. (Chairman) 94
Rutskoy, A. (former Vice-President) 28, 139, 140
Sacks, J, (former Harvard adviser to Russia) 139
Saratov (see *Saratov)*
Sevastopol 177
Shafranik, Y. (Minister of Fuel and Energy) 161
Shevardnadze, E. (FSU Foreign Minister, Georgia's KGB Chief, First Secretary of Communist Party) 75, 84
coup d'etat in Georgia 75
shipping problems 163-164, 166
Shnurnov, V. (President of KEDR) 130
Shumeyko, V. (Deputy Prime Minister of Russia, Speaker for Chamber of Russian Parliament, Vice-Premier Minister) 28, 135
Siberia (see *Siberia)*
socialistic nations 178
Stalin, J. 47-48, 72, 73, 76, 93-94, 178
creation of republics 178
education 76
Ossetia 76
Tatarstan 93-94
Stalin, Svetlana Alilueva (J. Stalin's daughter) 145
State Investment Committee 129
Stolypin (Prime Minister) 144
Suvorov, Generalissimus A. 151, 152
Tabakova, E. (Chief Engineer, Technical Department's Port Authority of Tuapse) 179
Taslitsky, E. (Deputy Minister of Fuel and Energy) 168
Tatarstan (see *Tatarstan)*
theft of resources 180-181, 193
government intervention 181
Rule 2-23 181, 193
Tolstoy, M., Ph.D. (member of Russian Parliament and of State Duma) 172
trade and investment 119-120, 1130-131, 134-135, 167, 170, 171, 181, 191, 193
joint ventures with the United States 119-120, 193
Rule 2-20 167, 191
Rule 2-21 170, 191
Rule 2-22 171
Rule 2-23 181, 193
laws concerning 130-131
Rule 2-6 130
political influences 134-135
Trans-Siberian Railway 159
travel problems 163-164
Udod, V. (Deputy Director of Regional State Tax Management branch in Tyumen) 174
Urazjtcev (member of Russian Parliament) 88-89
Uspenskaya, N. (Economics Advisor) 180-181
Rizjkov (Soviet Prime Minister) 180-181
Volgograd (formerly Stalingrad) 167
Western Siberia (see *Western Siberia)*
World War II 72, 73, 93-94
Yavlinsky (Economic Adviser, Member of Parliament) 125, 126

Yeltsin, B. (Chairman of Supreme Soviet of Russian Federation) 73, 89, 95, 97, 121, 124, 125, 128, 129, 132-133, 136-138, 142, 145, 152, 154-155, 173, 174, 177
Alekperov, Vagit (Lukopil President, Deputy Oil Minister) 142
Burburlis (Yeltsin associate) 128
condition of government 121
conflicts with Parliament 154-155
gold 173, 174
Golovaty, A. (Deputy Minister of Finance) 173
secret resolution creating state monopoly of reserves 173, 174
Udod, V. (Deputy Director of Regional State Tax Management branch in Tyumen) 174
democracy 136-138
property conflicts with Germany 152
Tatarstan oil agreement 97
Yermolov (General) 72
Ziuganov, G. (Leader of Russian Communist Party) 133, 161
Zjirinovsky (residential candidate) 145
Russian National-Patriotic Front *(Pamyat)* 94
Ostashvily, A. (Chairman) 94
Russians 3, 9, 12, 16, 40, 48-49, 71, 73
Rutskoy, A. (Vice-President) 28, 139, 140

S

Sacks, J. (former Harvard adviser to Russia) 139
Sadic-Khan, Orhan (Managing Director, Paine Webber Inc.) 96
safe behavior 67-68, 79, 112, 113
Rule 1-22 112
Rule 1-28 68, 112
Rule 1-30 79, 112
Sakhlainneft (Sakhalin Oil) 80
Galburaev, Ruslan (former Chief Engineer, Minister of Oil Industry) 74, 80, 83, 84-86, 91
Samarkand 47, 48, 49, 57, 113
accommodations 113
Saratov 18, 26, 28, 147-155
crime 151
influence of alcohol 151
Engels (former capital of the former Autonomous Republic of Volga's Germans) 151, 152
history 151, 152
Gorbachev, M. (Russian President) 147, 148-149
privatization of businesses 147, 148-149
vouchers in FSU businesses 148-149
history 151, 152
oil industry 147-155
contracts 147-148
Kazakhstan 147
Kirgizstan 147-148
Tajikistan 147-148
French exploration 154-155
negotiations 150-151
oil and gas 153-154
petroleum basins 153-154
privatization 147-149
vouchers in FSU businesses 148-149
reserves 153-154
Saratovneftegeofizika (Saratov Oil Geophysical Survey) 18, 147-149, 150, 155
Michurin, A. (President) 149
modernization 147
privatization 147-149
Zakharov, P. (Vice-President) 149-150, 155
Sarygulov, Dastan I. (Governor of Talas region, General Director of Kirgizstan's gold mining company) 44
Saudi Arabia 21
saving money 111
Rule 1-8 111
Rule 1-21 111
Rule 1-34 111

Sayidov (Chief of Dagestan Oil) 89-90
Schlumberger's oil equipment (the Chechen Republic) 84
Scotland 56
National Westminster Bank of Scotland 56
security 152-153, 193
Rule 2-17 153, 193
Rule 2-18 153, 193
seismic surveys 82
Bally (Rice University Professor) 82
Serebryansky, Anatoly T. (Deputy Director of gold mining company) 35
Sevastopol 177
Shafranik, Y. (Minister of Fuel and Energy) 161
Shamshiev, Bolot T. (Member of Kirgizstan Parliament) 12, 23, 33, 34
Sheblinka gas field (Ukraine) 182
Sheremetyevo (Moscow International Airport) 127
Shevardnadze, E. (FSU Foreign Minister, Georgia's KGB Chief, First Secretary of Communist Party) 75, 84
coup d'etat in Georgia 75
shipping problems 163-164, 166
Shnurov, V. (President, KEDR) 130
shopping 115
avoiding black markets 115
exchanging currency 115
follow hosts' advice 115
Shumeyko, V. (Deputy Prime Minister of Russia, Speaker for Chamber of Russian Parliament, Vice-Premier Minister) 28, 135
Siabeco Group (Swiss gold mining company) 17, 39, 45
Siberia 72, 74, 94-94, 103, 120, 133, 152, 155
American investments in 120
Krasnoyarsk region 120
oil industry 133
Omsk region 120
Tatarstan 93-94
Tyumen 152
Yakutia, Autonomous Republic of 103
prime minister 103
Siberian republic 55
Jakutiya 55
Sibnefteprovod (West Siberian Oil Pipelines) 167, 168-170, 171, 172
Chepursky, V. (Chief Engineer) 169-170, 171, 172
silver 57
Singapore 13
slavery 72
in the Chechen Republic 72
Slavyanskara (Moscow hotel) 127
Slovo Kirgizstana (Kirgizstan periodical) 32, 44, 46
Snegur, M. (President of Moldova) 28
socialistic nations 178
Sokolovsky, E. V. (Director, North Caucasian Petroleum Research Institute) 73-74, 84-86
Solotukhin (President of Intersystems, Uzbekistan) 48-49
Somalia 125
Sorokin, N. (Reporter for Voice of America) 12
South Korea 166
Soyuzalmazzoloto (Union's Diamonds and Gold enterprise) 16
Soviet Union (see *FSU)*
specialists, appointing local nationals (Rule 1-31) 79, 111
Sry-Durya River (Uzbekistan) 47, 49
Stalin, J. 47-48, 72, 73, 76, 93-94, 178
creation of republics 178
education 76
Ossetia 76
Tatarstan 93-94
Stalin, Svetlana Alilueva (J. Stalin's daughter) 145
Standard Oil 141
starting a business (Rule 1-36) 97, 110
Stavropol (Chechen Republic) 74, 81
Surkhan-Darya Depression (Uzbekistan) 49
Suvorov, Generalissimus A. 151, 152
Switzerland 13, 17, 39, 45
Siabeco Group (gold mining company) 17, 39, 45

T

Tabakova, E. (Chief, Technical Department's Port Authority of Tuapse) 179
Tajikistan 47
  arable land 47
Talas region (Kirgizstan) 44
  Sarygulov, Dastan I. (Governor of Talas region) 44
Taldy-Bulak gold deposit 14, 32, 38
Tamara (Princess of Georgia) 76
Tashkent (capital of Uzbekistan) 19, 29, 48, 49, 53, 58, 61, 62, 64, 67, 68, 113
  accommodations 113
  crime in 68-69, 113
Taslitsky (Deputy Minister of Fuel and Energy) 168
Tatarstan
  Crimea, Autonomous Republic of 93-94
    Stalin, J. 93-94
    World War II 93-94
  Dautov, Marsel A. (Minister of Community Services) 99-103
  government 96
  history 93-94
  industries 95, 99-102, 133
    chemical 95
    manufacturing 95
    oil industry
      agreement with Russia 97
      government involvement 96
      production rates 95
      refineries 95, 99-102
    secondary and tertiary oil resources 95
    Devonian and Carboniferous deposits 95
    limestones and sandstones 95
    Permian and Triassic sequences 95-96
      solid bitumen 95-96, 104
      sulfur content 96, 99
    Oil Ministry in Moscow 96
    Rijanov, A. (Chief gas and oil expert of Cabinet of Ministers) 96, 98, 99-102
    Romashkinskoe oil fields 97
    Shaymiev (President) 97, 98
      Russian oil agreement 97
  Islam 93, 102
  Ivan the Terrible 93
  Kaganovich, L. (First Secretary in Kiev) 94
  Kazan (capital) 93, 96, 98, 99, 102, 113
    accommodations 113
  khan (title) 93
  Minister of Yakutia 103
  nationalistic movement 95
  relations with FSU and Russian Parliaments 103-104
  Rijanov, A. (Chief gas and oil expert of Cabinet of Ministers) 96, 98, 99-102
  Sadic-Khan, Orhan (Managing Director, Paine Webber Inc.) 96
  Shaymiev (President) 97, 98
    oil agreement with Russia 97
  Tatars 93-94
    Heroes of the Soviet Union 93-94
    Moslems 93
    population 93
  theft of oil 181
  Viapaerking (subsidiary of Global Natural Resources) 169
  Yeltsin, B. (Chairman of Supreme Soviet of Russian Federation) 95, 97
    oil agreement 97
Tavorovsky, V. (General Director of Uzbek Oil and Gas Pipelines Installation) 49
Taganrog 177
Tbilisi 75, 76
  Babilashvily, A. (Professor of History, University of Tbilisi) 76
Tekenov, S. (General Director of Kirgizstan State Committee of Geological Exploration) 19, 22, 30-34, 40-43
Tengizchevroil 20
  Dupre, M. (general director) 20
Teutonic culture 72
Texaco 120, 135
  Komi oil project 120, 135

Timan-Pechora 135
Thatcher, Margaret (British Petroleum Representative, Prime Minister) 141
Azerbaijan oil deal 141
Tien Shan mountains 9, 53
Timan-Pechora (American-owned Russian oil company) 135
tin 14
Tobago 142
oil fields 142
Tobolsk 169, 170
Tolstoy, M., Ph.D. (member of Russian Parliament and State Duma) 172
trade and investment in Russia 119-120, 1130-131, 134-135, 167, 170, 171, 181, 191, 193
joint ventures with the United States 119-120, 193
Rule 2-20 167, 191
Rule 2-21 170, 191
Rule 2-22 171
Rule 2-23 181, 193
laws concerning 130-131
Rule 2-6 130
political influences 134-135
Trans-Siberian Railway 159
travel 39, 53, 79, 105-106, 112, 163-164
air 53
problems 163-164
Rule 1-14 39, 105-106
Rule 1-22 53, 112
Rule 1-30 79, 112
Trinidad 142
oil fields 142
Tsars 177-178
Tuapse 179, 180
tungsten 14
Tunis 142
oil fields 142
Turkey 12, 19, 55, 76, 101, 102, 141, 149, 178, 179, 181
Black Sea 179
BMB (gold company) 55
Istanbul 19, 76, 102, 181
Blue Mosque 102
pipelines 141
Turkmenistan
Ataberdiev, M. (Deputy Minister of Fuel) 67
Chardzjou refinery 66
education 78
Krasnovodsk oil refinery 66
natural gas 67
recoverable reserves 67
Niyazov (President) 66
oil industry 66-67
Chardzjou refinery 66
Krasnovodsk refinery 66
railroad to Iran 67
railroad to Iran 67
Tursunov, B. (Chancellor, First Uzbek Independent University of Business and Diplomacy) 48
Tyumen (Western Siberia) 152, 158, 161, 167, 170, 171

U

Udod, V. (Deputy Director of Regional State Tax Management branch in Tyumen) 174
unfranchised status, strategy of 130, 135, 138, 140, 192
2-6 130
2-9 135
2-10 138
2-13 140
Ukraine
Black Sea 179
foreign investments 182-183
history 178
Kish, W. (Seagram Manager) 182
Korchinsky, D (Deputy Chairman, National Assembly) 179
nationalism 142, 187-188
Alexey, Tsar 178
Crimea 187-188
oil industry
description of 182
Dnieper-Donetsk Depression 182
gas fields 182
offshore zone 184
Pre-Carpathian Fordeep 182, 184
Ukrgasprom (Ukraine's Gas Industry) 183
population 94
selling food to Moscow 187-188
Sheblinka gas field 182

trade with Uzbekistan 67
underground mining (gold) 58
United Parcel Service 114
United States
affirmative actions programs 81
Alaska 20
America 6, 12, 21, 29, 43
American diplomacy 6
American Federal Reserve System to the Byelorus Republic 143
Laws, R. (Representative) 143
American trade and investments in Russia 119-120
Ames (CIA Officer and Russian Spy) 75-76
Astrakhan oil deal 134-135
Baker, J. (Secretary of State) 7, 8
Bally, (Professor, Rice University) 82
seismic surveys 82
Brown, R. H. (Secretary of Commerce) 178
Bush, George 43, 122
Christopher, Warren (Secretary of State) 6, 7
CIA 75-76
Ames (CIA Officer and Russian Spy) 75-76
Clinton, William 122
Colorado 41, 64, 88, 103, 141-142
Apache (oil company) 141-142
Denver 103, 141-142
Romer, Roy (Governor) 64, 88
Columbia University 19
Constitution 143
Stolypin (Russian Prime Minister) 143
courts 142
Frankenburgh-Lukoil law suit 142
Embassy in Uzbekistan 64
First Secretary of the U.S. Embassy in Kirgizstan 33
Gore, Al 8
government negotiations with the FSU 174
Hutson, T. R. (First Secretary of Embassy in Bishkek, Kirgizstan) 6
Johnson, Lyndon B. 81
Kennedy, John F. 81
Kissinger, H. (former Secretary of State) 122
Marines, U.S. 88
Matlock, J. (former Ambassador to the FSU) 33
Michigan University's Center of the Study of the U.S.S.R. and East Europe 140
New York 19, 73
crime 73
New York City 72
Norsk Hydro 135
overdependence on Mideastern oil producers 176
political conditions in the Chechen Republic 75-76
U.S. involvement 75-76
Ames, (CIA Officer and Russian Spy) 75-76
Prudhoe Bay 20
Rice University 82
Bally, (Professor) 82
seismic surveys 82
Rocky Mountains 41
trade and investment with Russia 119-120, 121-122
joint ventures 119-120
perception of economic and political situations 121-122
Gallup poll 121
reasons for different business situations 121
Vietnam 74
Voice of America 12
Washington, D.C. 134
University of Tbilisi 76
Babilashvily, A. (Professor of History) 76
uranium 14
Urazjtcev (member of the first Russian Parliament) 88-89
Uspenskaya, N. (Economics Advisor) 180-181
Rizjkov (Prime Minister) 180-181
Uzbekistan
Almalyk refining plant (gold) 58
Almay Say Fault Zone 58
Angren Mill (gold) 58

alphabets 48
Amu-Darya River 47
Andropov, Yuri (KGB head, General Secretary of the FSU Communist Party) 52
Aral Sea 52
Bokhara-Hivinsky basin 50
business management 69
cities and regions
    Bokhara 47, 49, 113
        accommodations 113
    Hiva 47
    Samarkand 47, 48, 49 57, 113
        accommodations 113
    Tashkent (capital) 19, 29, 48, 49, 53, 58, 61, 62, 64, 67, 68, 113
        accommodations 113
        crime 68-69, 113
Communism 47, 52-53
    Rashidov, S. (First Secretary of Central Committee of Communist Party) 52-53
cotton 50, 52
Darius (Shah) 47
Denau train station 56
Dushanbe-Termez road 56
education 78
Fergana intermountain basin 48-49, 50, 53, 62-63
    oil drilling 50
    oil-well explosion 62-63
First Uzbek Independent University of Business and Diplomacy 48
    Tursunov, B. (Chancellor) 48
gas 48, 50, 52
gold
    Almalyk refining plant 58
    Angren Mill 58
    BMB (Turkish gold company) 55
    EBRD 55
    deposits 53, 55, 57, 58, 60
        Akgol 58
        Jalair 58
        Karakutan 58
        Kattaich 58
        Kayragash 58
        Keskan 58
        Koch Bulak 58
        Kyzyl Almasay 56, 58
            mining method (diagram) 60
        Mezjdurchye 58
        Muruntau 53
        Samarchuk 58
        Samich 58
        Sop 58
        Temirkabuk 58
        Uchkulach 58
        Yakhton 58
        Zarmitan 55, 57
    heap leaching technology 56
    Hoshars (party for neighbors, Uzbekistan) 63-66
        conducting business at 63-66
    Intersystems 48-49
        Solotukhin, V. (President) 48-49
    London's gold market 62
    maps
        eastern Uzbekistan's deposits 59
        Uzbekistan's deposits 54
    Margjanbulack Enriching Plant 57
    mineable reserves 58
    Newmont (North American gold company) 17, 29, 37, 39, 53, 55, 64
        Cambre, M. (CEO) 55
            European Development Bank 55
    privatization of industry 61-62, 148-149
        vouchers in FSU businesses 148-149
    refining plants 58
        Almalyk 58
    resources 58
    underground mining 58
Gorbachev 48
Gorbachev, M. (Deputy Minister of Interior Affairs, L. Brezhnev's son-in-law) 52-53
healthcare 52
heap leaching technology 56
Hiva 47
Hoshars (party for neighbors, Uzbekistan) 63-66

conducting business at 63-66
Islam 47, 48
Karimov (President) 61
Karshy basin 50
Khackulov (Deputy Prime Minister, Chairman of *Uzbekeftegaz)* 56, 63
Khandiza polymetallic deposit 56
Kyzyl-Kum desert 47, 49, 53
languages 47
Lower Paleozoic sediments 58
Margjanbulack Enriching Plant 57
mineral resources 53, 57, 58
cadmium 57
copper 57
gold 53, 58
lead 57
silver 57
zinc 57
Minister of Foreign Trade 49
Muruntau gold deposit 53
Najimov, S. (Chairman, State Committee for Precious Metals) 61, 64
National Westminster Bank of Scotland (NatWest) 56
oil industry 49, 50, 149
Bokhara-Hivinsky basin 50
drilling 50
Gumkhana 50
Kubarnalakma 49, 50
pipeline 50
Surkhan-Darya Depression 49, 50
Ust-Yurt basin 50
Oil Ministry in Moscow 96
Osh 48
Pan-Turkism 48
perestroyka 52
Persian Gulf 67
population 47
pollution 52
Aral Sea 52
privatization 61-62
Rakhimov, A. (First Deputy and Chairman, Uzbek Oil and Gas) 62-63, 64
Rashidov, S. (First Secretary of Central Committee of Communist Party) 52-53
Russians 48-49
Stalin, J. 47-48
state TV 69
Surkhan-Darya Depression 49
Kubamalakma oil field 49
Syr-Darya River 47, 49
Tavorovsky, V. (General Director of Uzbek Oil and Gas Pipelines Installation) 49
Tien Shan mountains 9, 53
Tursunov, B. (Chancellor, First Uzbek Independent University of Business and Diplomacy) 48
Ust-Yurt basin 50
Uzbekistan-Conctact 56
Kononova, M. (Editor) 56
Uzbeks 47
Sunni Muslims 47
Shiite Muslims 47
Tajiks 47
Ishmaelites 47
Uzuh train station 56

V

vendetta 71
the Chechen Republic 71
Vietnam 74
visiting the FSU 113-115
climate 115
communications 114
courtesy 115
entertainment 114
exchanging currency 115
hotels 113
meals 114
shopping 115
Volga-Don Channel 181
Volga River 18, 151, 152, 155, 181
Volga-Ural Basin oil fields 82, 153-154
Volgograd (formerly Stalingrad) 167
Voltaire 177
von Bismarck, O. (Chancellor of the German Empire) 123

W

wealth, expressing (Rule 1-14) 39, 105-106

Western Siberia
Autonomous Republic of Khanty and Mancy 160
Balyakov, V. (Senior Geologist, Samotlorneft) 161
Bashkirya 159
Bratsk 163
Chepursky, V. (Chief Engineer of West Siberian Oil Pipelines) 169-170
Irkutsk 162, 163
oil industry
areas where western investment is encouraged 162-168
description 157-161
economic and political pressures 159-161
geography and geology 157-160, 1164-165
production and reserves 157-160
ecological damage 167
fraud in cleanups 167
expenses 164-166
production 165
shipping 166
exploration 175
with submarines 175
local oil operating companies 160
autonomy from government 160
sale of oil (legal and illegal) 160
oil and gas fields 158, 159, 164
Berezov 158
Chonsky, V. 164
Markovsky 164
Shaim 158
Urengoy 159
Yamburg 159
Yarakhtinsky 164
pipelines 158, 159, 164, 169, 175
construction of 164
Druzhba 169
exploration of 159
pollution 168, 169
refineries 159, 168, 169
European Russian 168
Moscow 168
Nizhny 168
Novgorod 168
Ryazan 168
Yaroslav 168
Russian 159, 169
Achinsk 159
Omsk 159, 169
Tomsk 159, 169
Volga-Ural Refineries 169
Ishimbay 169
Samara 169
Saratov 169
Syzran 169
Rule 2-19 166, 191
Sibnefteprovod (West Siberian Oil Pipelines) 167, 168-170, 171, 172
Tobolsk 169, 170
Tyumen 152, 158, 161, 167, 170, 171
Udod, V. (Deputy Director, Regional State Tax Management branch, Tyumen) 174
Yamal Nenetz Autonomous District 160
willingness to help (Rule 1-23) 61, 106
women 35, 67-68, 108, 115
Rule 1-11 35, 108
Rule 1-26 108
Rule 1-28 68, 112
World Bank 19, 142, 161

Y

Yakutia, Autonomous Republic of 103
prime minister 103
Yamal Nenetz Autonomous District (Western Siberia) 160
Yavlinsky (Economic Adviser, Member of Parliament) 125, 126
Yeltsin, B. (Chairman of Supreme Soviet of Russian Federation) 73, 89, 95, 97, 121, 124, 125, 128, 129, 132-133, 136-138, 142, 145, 152, 154-155, 173, 174, 177

Alekperov, Vagit (Lukopil President, Deputy Oil Minister) 142
Burburlis (Yeltsin associate) 128
condition of government 121
conflicts with Parliament 154-155
gold 173, 174
Golovaty, A. (Deputy Minister of Finance) 173
secret resolution creating state monopoly of reserves 173, 174
democracy 136-138, 192
property conflicts with Germany 152
Tatarstan oil agreement 97
Yermolov (Russian General) 72
Yuzjmorneftegeofizika (Southern Marine Geophysics) 187

Z

Zakharov, P. (Vice-President, Saratov Oil Geophysical Survey) 149-150, 155
Zarmitan gold deposit 55, 57
Zenina, Oksana M. (European Representative, Dunavant Cotton & Ginning Service) 68
Zhirinovsky 74
zinc 57
Ziuganov, G. (Chairman of Russian Communist Party) 133, 161
Zjirinovsky (Russian presidential candidate) 145
Zjumagulov, A. (Governor of the Chu Region, Kirgizstan) 45
Zolotukhin, V. (President of Intersystems, Member of the Parliament) 48-49